TEACHER GUIDE | 1 Credit | Weekly Lesson Schedule

9th–12th Grade | Scie

Life Science: Origins & Scientific Theory

Author: Dr. Carl Werner

Master Books Creative Team:

Editor: Craig Froman

Design: Terry White

Cover Design: Diana Bogardus

Copy Editors:
Judy Lewis
Willow Meek

Curriculum Review:
Kristen Pratt
Laura Welch
Diana Bogardus

First printing: March 2013
Fifth printing: August 2020

Master Books®, P.O. Box 726, Green Forest, AR 72638

Master Books® is a division of the New Leaf Publishing Group, Inc.

ISBN: 978-1-68344-115-1
ISBN: 978-1-61458-659-3 (digital)

Unless otherwise noted, Scripture quotations are from the New King James Version of the Bible.

Printed in the United States of America

Please visit our website for other great titles:
www.masterbooks.com

About the Author:

In his sophomore year of college, **Dr. Carl Werner** was challenged by a fellow classmate with these words: "I bet you can't prove evolution." This began Dr. Werner's quest for an answer. After 18 years of study, Dr. Werner would begin travelling to the best museums and dig sites around the globe, photographing thousands of original fossils and the actual fossil layers where they were found, and interviewing scientists on the issue. After his years of study and the evidence he has seen, Dr. Werner realized he had reached a most unexpected truth – the truth of a biblical creation.

Your reputation as a publisher is stellar. It is a blessing knowing anything I purchase from you is going to be worth every penny!

—Cheri ★★★★★

Last year we found Master Books and it has made a HUGE difference.

—Melanie ★★★★★

We love Master Books and the way it's set up for easy planning!

—Melissa ★★★★★

You have done a great job. MASTER BOOKS ROCKS!

—Stephanie ★★★★★

Physically high-quality, Biblically faithful, and well-written.

—Danika ★★★★★

Best books ever. Their illustrations are captivating and content amazing!

—Kathy ★★★★★

Affordable
Flexible
Faith Building

Table of Contents

Using This Teacher Guide

Overview: This *Life Science: Origins & Scientific Theory Teacher Guide* contains materials for use with *Evolution: The Grand Experiment Vol. 1* and *Living Fossils*. By developing a deeper understanding of these concepts, students will be able to develop and support a strong worldview.

Course Description: This course is intended to help a student assess information about evolution and creation, and based on the information provided for each, form his or her own understanding of this issue. The author spent 30 years in a challenge to prove evolution, yet the more he learned, the more the truth of God's Word became apparent in the evidence he found while traveling the world viewing artifacts and speaking to scholars and museum officials.

Upon completion of this course, students will have a thorough understanding of the theory of evolution and its limits. Students will develop scientific critical thinking skills through careful analysis of evidence and comparing the merits of different theories. Students will study paleontology, biology, and geology as they relate to the study of origins through an exploration of living fossils.

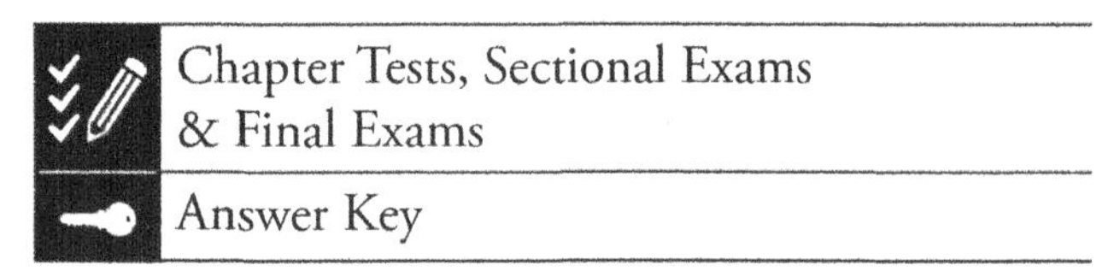

Workflow:
Step 1: Teacher leads Discussion Questions.
Step 2: Student watches DVD (if applicable).
Step 3: Student reads chapter.
Step 4: Student is given and completes Chapter Worksheet Questions after reading the chapter.
Step 5: Teacher administers Chapter Test.
Step 6: Teacher administers Sectional Exams where indicated.
Step 7: Teacher administers Comprehensive Final Exams where indicated.

Lesson Scheduling: Space is given for assignment dates. There is flexibility in scheduling. For example, the parent may opt for a M–W schedule rather than a M, W, F schedule. Each week listed has five days, but, due to vacations, the school work week may not be M–F. Adapt the days to your school schedule. As the student completes each assignment, he/she should put an X in the box.

How to Use this Course

You will follow the course calendar whether you are using the course for one or multiple students.

Chapter Discussion Questions: Teachers are encouraged to participate with the student as they complete the discussion questions. The purpose of the Chapter Purpose section is to introduce the chapter to the student. The Discussion Questions are meant to be thought provoking. The student may not know the answers, but should answer with their thoughts, ideas, and knowledge of the subject using sound reasoning and logic. They should study the answers and compare them with their own thoughts. We recommend the teacher discuss the questions, the student's answers, and the correct answers with the student. This section should not be used for grading purposes.

DVD: Each DVD is watched in its entirety to familiarize the student with each book in the course. They will watch it again as a summary when they complete each book. Students may also use the DVD for review, as needed, as they complete the course.

Chapter Worksheets: The worksheets are foundational to helping the student learn the material and come to a deeper understanding of the concepts presented. Often, the student will compare and contrast what we should find in the fossil record and in living creatures if evolution were true with what we actually find. This comparison clearly shows evolution is an empty theory simply based on the evidence. God's Word can be trusted and displayed both in the fossil record and in what we see in living animals.

Tests and Exams: There is a test for each chapter, sectional exams, and a comprehensive final exam for each book.

First Semester Suggested Daily Schedule

Part 1 *Evolution: The Grand Experiment: The Quest for an Answer*

Date	Day	Assignment	Due Date	✓	Grade
		First Semester — First Quarter			
Week 1	Day 1	Watch DVD in entirety to become familiar with course topics			
	Day 2	Complete Chapter 1 Discussion Questions • Pages 17-18 *Teacher Guide* • (TG) Chapter 1: The Origin of Life: Two Opposing Views Read pages 1-10 • *Evolution: The Grand Experiment* • (ETGE)			
	Day 3	Complete Chapter 1 Worksheet • Pages 19-20 • (TG)			
	Day 4	Take **Chapter 1 Test** • Pages 223-224 • (TG)			
	Day 5	Complete Chapter 2 Discussion Questions • Pages 21-22 • (TG) Chapter 2: Evolution's False Start: Spontaneous Generation Read pages 11-22 • (ETGE)			
Week 2	Day 6	Complete Chapter 2 Worksheet • Pages 23-24 • (TG)			
	Day 7	Take **Chapter 2 Test** • Pages 225-226 • (TG)			
	Day 8	Complete Chapter 3 Discussion Questions • Pages 25-26 • (TG) Chapter 3: Darwin's False Mechanism for Evolution: Acquired ... Read pages 23-30 • (ETGE)			
	Day 9	Complete Chapter 3 Worksheet • Pages 27-28 • (TG)			
	Day 10	Take **Chapter 3 Test** • Pages 227-228 • (TG)			
Week 3	Day 11	Study Day-Chapters 1-3			
	Day 12	Study Day-Chapters 1-3			
	Day 13	Take **Sectional Exam 1** Chapters 1-3 • Pages 315-316 • (TG)			
	Day 14	Complete Chapter 4 Discussion Questions • Pages 29-30 • (TG) Chapter 4: Natural Selection and Chance Mutations Read pages 31-37 • (ETGE)			
	Day 15	Chapter 4: Natural Selection and Chance Mutations Read pages 38-54 • (ETGE)			
Week 4	Day 16	Complete Chapter 4 Worksheet: Questions 1-13 Pages 31-32 • (TG)			
	Day 17	Complete Chapter 4 Worksheet: Questions 14-27 Pages 33-34 • (TG)			
	Day 18	Take **Chapter 4 Test** • Pages 229-230 • (TG)			
	Day 19	Complete Chapter 5 Discussion Questions • Pages 35-36 • (TG) Chapter 5: Similarities: A Basic Proof of Evolution? Read pages 55-59 • (ETGE)			
	Day 20	Chapter 5: Natural Selection and Chance Mutations Read pages 60-72 • (ETGE)			

Date	Day	Assignment	Due Date	✓	Grade
Week 5	Day 21	Complete Chapter 5 Worksheet: Questions 1-13 Pages 37-38 • (TG)			
	Day 22	Complete Chapter 5 Worksheet: Questions 14-22 Pages 39-40 • (TG)			
	Day 23	Take **Chapter 5 Test** • Pages 231-232 • (TG)			
	Day 24	Complete Chapter 6 Discussion Questions • Pages 41-42 • (TG) Chapter 6: The Fossil Record and Darwin's Prediction Read pages 73-86 • (ETGE)			
	Day 25	Complete Chapter 6 Worksheet • Pages 43-44 • (TG)			
Week 6	Day 26	Take **Chapter 6 Test** • Page 233 • (TG)			
	Day 27	Complete Chapter 7 Discussion Questions • Pages 45-46 • (TG) Chapter 7: The Fossil Record of Invertebrates Read pages 87-94 • (ETGE)			
	Day 28	Complete Chapter 7 Worksheet • Pages 47-48 • (TG)			
	Day 29	Take **Chapter 7 Test** • Page 235 • (TG)			
	Day 30	Complete Chapter 8 Discussion Questions • Pages 49-50 • (TG) Chapter 8: The Fossil Record of Fish • Read pages 95-98 • (ETGE)			
Week 7	Day 31	Complete Chapter 8 Worksheet • Pages 51-52 • (TG)			
	Day 32	Take **Chapter 8 Test** • Page 237 • (TG)			
	Day 33	Complete Chapter 9 Discussion Questions • Pages 53-54 • (TG) Chapter 9: The Fossil Record of Bats • Read pages 99-104			
	Day 34	Complete Chapter 9 Worksheet • Pages 55-56 • (TG)			
	Day 35	Take **Chapter 9 Test** • Page 239 • (TG)			
Week 8	Day 36	Study Day-Chapters 4-9			
	Day 37	Study Day-Chapters 4-9			
	Day 38	Study Day-Chapters 4-9			
	Day 39	Take **Sectional Exam 2** Chapters 4-9 • Pages 317-320 • (TG)			
	Day 40	Complete Chapter 10 Discussion Questions • Pages 57-58 • (TG) Chapter 10: The Fossil Record of Pinnipeds: Seals and Sea Lions Read pages 105-112 • (ETGE)			
Week 9	Day 41	Complete Chapter 10 Worksheet • Pages 59-62 • (TG)			
	Day 42	Take **Chapter 10 Test** • Page 241 • (TG)			
	Day 43	Complete Chapter 11 Discussion Questions • Pages 63-64 • (TG) Chapter 11: The Fossil Record of Flying Reptiles Read pages 113-116 • (ETGE)			
	Day 44	Complete Chapter 11 Worksheet • Pages 65-66 • (TG)			
	Day 45	Take **Chapter 11 Test** • Page 243 • (TG)			

Date	Day	Assignment	Due Date	✓	Grade
First Semester — Second Quarter					
Week 1	Day 46	Complete Chapter 12 Discussion Questions • Pages 67-68 • (TG) Chapter 12: The Fossil Record of Dinosaurs Read pages 117-128 • (ETGE)			
	Day 47	Complete Chapter 12 Worksheet: Questions 1-13 Pages 69-70 • (TG)			
	Day 48	Complete Chapter 12 Worksheet: Questions 14-25 Pages 71-72 • (TG)			
	Day 49	Take **Chapter 12 Test** • Page 245 • (TG)			
	Day 50	Complete Chapter 13 Discussion Questions • Pages 73-74 • (TG) Chapter 13: The Fossil Record of Whales Read pages 129-146 • (ETGE)			
Week 2	Day 51	Complete Chapter 13 Worksheet 1 • Pages 75-78 • (TG)			
	Day 52	Complete Chapter 13 Worksheet 2 • Pages 79-80 • (TG)			
	Day 53	Take **Chapter 13 Test** • Page 247 • (TG)			
	Day 54	Complete Chapter 14 Discussion Questions • Pages 81-82 • (TG) Chapter 14: The Fossil Record of Birds Part 1: *Archaeopteryx* Read pages 147-164 • (ETGE)			
	Day 55	Complete Chapter 14 Worksheet 1 • Pages 83-84 • (TG)			
Week 3	Day 56	Complete Chapter 14 Worksheet 2 • Pages 85-86 • (TG)			
	Day 57	Take **Chapter 14 Test** • Page 249 • (TG)			
	Day 58	Complete Chapter 15 Discussion Questions • Pages 87-88 • (TG) Chapter 15: The Fossil Record of Birds Part 2: Feathered Dinosaurs Read pages 165-184 • (ETGE)			
	Day 59	Complete Chapter 15 Worksheet 1 • Pages 89-90 • (TG)			
	Day 60	Complete Chapter 15 Worksheet 2 • Pages 91-92 • (TG)			
Week 4	Day 61	Take **Chapter 15 Test** • Page 251 • (TG)			
	Day 62	Study Day-Chapters 10-15			
	Day 63	Study Day-Chapters 10-15			
	Day 64	Study Day-Chapters 10-15			
	Day 65	Take **Sectional Exam 3** Chapters 10-15 • Pages 321-324 • (TG)			
Week 5	Day 66	Complete Chapter 16 Discussion Questions • Pages 93-94 • (TG) Chapter 16: The Fossil Record of Flowering Plants Read pages 185-190 • (ETGE)			
	Day 67	Complete Chapter 16 Worksheet • Pages 95-96 • (TG)			
	Day 68	Take **Chapter 16 Test** • Page 253 • (TG)			
	Day 69	Complete Chapter 17 Discussion Questions • Pages 97-98 • (TG) Chapter 17: The Origin of Life — Part 1: The Formation of DNA Read pages 191-198 • (ETGE)			
	Day 70	Complete Chapter 17 Worksheet 1 • Pages 99-100 • (TG)			

Date	Day	Assignment	Due Date	✓	Grade
Week 6	Day 71	Complete Chapter 17 Worksheet 2 • Pages 101-102 • (TG)			
	Day 72	Take **Chapter 17 Test** • Page 255 • (TG)			
	Day 73	Complete Chapter 18 Discussion Questions Pages 103-104 • (TG) Chapter 18: The Origin of Life — Part 2: The Formation of Proteins • Read pages 199-204 • (ETGE)			
	Day 74	Complete Chapter 18 Worksheet • Pages 105-106 • (TG)			
	Day 75	Take **Chapter 18 Test** • Page 257 • (TG)			
Week 7	Day 76	Complete Chapter 19 Discussion Questions • Pages 107-108 • (TG) Chapter 19: The Origin of Life — Part 3: The Formation of Amino Acids • Read pages 205-210 • (ETGE)			
	Day 77	Complete Chapter 19 Worksheet 1 • Page 109 • (TG)			
	Day 78	Complete Chapter 19 Worksheet 2 • Pages 111-112 • (TG)			
	Day 79	Take **Chapter 19 Test** • Pages 259-260 • (TG)			
	Day 80	Study Day-Chapters 16-19			
Week 8	Day 81	Study Day-Chapters 16-19			
	Day 82	Study Day-Chapters 16-19			
	Day 83	Study Day-Chapters 16-19			
	Day 84	Take **Sectional Exam 4** Chapters 16-19 • Pages 325-328 • (TG)			
	Day 85	Watch DVD in its entirety as a summary.			
Week 9	Day 86	Study Day-Chapters 1-19			
	Day 87	Study Day-Chapters 1-19			
	Day 88	Study Day-Chapters 1-19			
	Day 89	Study Day-Chapters 1-19			
	Day 90	Take **Comprehensive Exam** Chapters 1-19 Pages 329-332 • (TG)			
		Mid-Term Grade			

Second Semester Suggested Daily Schedule

Part 2 *Evolution: The Grand Experiment: Living Fossils*

Date	Day	Assignment	Due Date	✓	Grade
		Second Semester — Third Quarter			
Week 1	Day 91	Complete Chapter 1 Discussion Questions • Pages 115-116 • (TG) Chapter 1: The Challenge That Would Change My Life Read pages 1-6 • (ETGE)			
	Day 92	Complete Chapter 1 Worksheet • Pages 117-118 • (TG)			
	Day 93	Take **Chapter 1 Test** • Page 263 • (TG)			
	Day 94	Complete Chapter 2 Discussion Questions • Pages 119-120 • (TG) Chapter 2: How Can You Verify Evolution? Read pages 7-14 • (ETGE)			
	Day 95	Complete Chapter 2 Worksheet • Pages 121-122 • (TG)			
Week 2	Day 96	Take **Chapter 2 Test** • Page 265 • (TG)			
	Day 97	Complete Chapter 3 Discussion Questions • Pages 123-124 • (TG) Chapter 3: The Naming Game • Read pages 15-28 • (ETGE)			
	Day 98	Complete Chapter 3 Worksheet • Pages 125-127 • (TG)			
	Day 99	Take **Chapter 3 Test** • Page 267 • (TG)			
	Day 100	Complete Chapter 4 Discussion Questions • Pages 129-130 • (TG) Chapter 4: Echinoderms • Read pages 29-44 • (ETGE)			
Week 3	Day 101	Complete Chapter 4 Worksheet • Pages 131-132 • (TG)			
	Day 102	Take **Chapter 4 Test** • Page 269 • (TG)			
	Day 103	Complete Chapter 5 Discussion Questions • Pages 133-134 • (TG) Chapter 5: Aquatic Arthropods • Read pages 45-56 • (ETGE)			
	Day 104	Complete Chapter 5 Worksheet • Pages 135-136 • (TG)			
	Day 105	Take **Chapter 5 Test** • Page 271 • (TG)			
Week 4	Day 106	Complete Chapter 6 Discussion Questions • Pages 137-138 • (TG) Chapter 6: Land Arthropods • Read pages 57-70 • (ETGE)			
	Day 107	Complete Chapter 6 Worksheet • Pages 139-140 • (TG)			
	Day 108	Take **Chapter 6 Test** • Page 273 • (TG)			
	Day 109	Complete Chapter 7 Discussion Questions • Pages 141-142 • (TG) Chapter 7: Bivalve Shellfish • Read pages 71-76 • (ETGE)			
	Day 110	Complete Chapter 7 Worksheet • Pages 143-144 • (TG)			
Week 5	Day 111	Take **Chapter 7 Test** • Page 275 • (TG)			
	Day 112	Complete Chapter 8 Discussion Questions • Pages 145-146 • (TG) Chapter 8: Snails • Read pages 77-82 • (ETGE)			
	Day 113	Complete Chapter 8 Worksheet • Pages 147-148 • (TG)			
	Day 114	Take **Chapter 8 Test** • Page 277 • (TG)			
	Day 115	Complete Chapter 9 Discussion Questions • Pages 149-150 • (TG) Chapter 9: Other Types of Shellfish • Read pages 83-88 • (ETGE)			

Date	Day	Assignment	Due Date	✓	Grade
	Day 116	Complete Chapter 9 Worksheet • Pages 151-152 • (TG)			
	Day 117	Take **Chapter 9 Test** • Page 279 • (TG)			
Week 6	Day 118	Complete Chapter 10 Discussion Questions • Pages 153-156 • (TG) Chapter 10: Worms • Read pages 89-92 • (ETGE)			
	Day 119	Complete Chapter 10 Worksheet • Page 157 • (TG)			
	Day 120	Take **Chapter 10 Test** • Page 281 • (TG)			
	Day 121	Complete Chapter 11 Discussion Questions Pages 159-160 • (TG) Chapter 11: Sponges and Corals • Read pages 93-98 • (ETGE)			
Week 7	Day 122	Complete Chapter 11 Worksheet • Page 161 • (TG)			
	Day 123	Take **Chapter 11 Test** • Page 283 • (TG)			
	Day 124	Study Day-Chapters 1-11			
	Day 125	Study Day-Chapters 1-11			
	Day 126	Study Day-Chapters 1-11			
	Day 127	Take **Sectional Exam 1** Chapters 1-11 • Pages 335-338 • (TG)			
Week 8	Day 128	Complete Chapter 12 Discussion Questions • Pages 163-164 • (TG) Chapter 12: Bony Fish • Read pages 99-116 • (ETGE)			
	Day 129	Complete Chapter 12 Worksheet • Pages 165-166 • (TG)			
	Day 130	Take **Chapter 12 Test** • Page 285 • (TG)			
	Day 131	Complete Chapter 13 Discussion Questions • Pages 167-168 • (TG) Chapter 13: Cartilaginous Fish • Read pages 117-124 • (ETGE)			
	Day 132	Complete Chapter 13 Worksheet • Pages 169-170 • (TG)			
Week 9	Day 133	Take **Chapter 13 Test** • Page 287 • (TG)			
	Day 134	Complete Chapter 14 Discussion Questions • Pages 171-172 • (TG) Chapter 14: Jawless Fish • Read pages 125-128 • (ETGE)			
	Day 135	Complete Chapter 14 Worksheet • Page 173 • (TG)			
		Second Semester — Fourth Quarter			
	Day 136	Take **Chapter 14 Test** • Page 289 • (TG)			
	Day 137	Complete Chapter 15 Discussion Questions • Pages 175-176 • (TG) Chapter 15: Amphibians • Read pages 129-136 • (ETGE)			
Week 1	Day 138	Complete Chapter 15 Worksheet • Page 177 • (TG)			
	Day 139	Take **Chapter 15 Test** • Page 291 • (TG)			
	Day 140	Complete Chapter 16 Discussion Questions • Pages 179-180 • (TG) Chapter 16: Crocodilians • Read pages 137-142 • (ETGE)			
	Day 141	Complete Chapter 16 Worksheet • Pages 181-182 • (TG)			
	Day 142	Take **Chapter 16 Test** • Page 293 • (TG)			
Week 2	Day 143	Complete Chapter 17 Discussion Questions • Pages 183-184 • (TG) Chapter 17: Snakes • Read pages 143-146 • (ETGE)			
	Day 144	Complete Chapter 17 Worksheet • Page 185 • (TG)			
	Day 145	Take **Chapter 17 Test** • Page 295 • (TG)			

Date	Day	Assignment	Due Date	✓	Grade
Week 3	Day 146	Complete Chapter 18 Discussion Questions • Pages 187-188 • (TG) Chapter 18: Lizards • Read pages 147-154 • (ETGE)			
	Day 147	Complete Chapter 18 Worksheet • Page 189 • (TG)			
	Day 148	Take **Chapter 18 Test** • Page 297 • (TG)			
	Day 149	Complete Chapter 19 Discussion Questions • Pages 191-192 • (TG) Chapter 19: Turtles • Read pages 155-160 • (ETGE)			
	Day 150	Complete Chapter 19 Worksheet • Pages 193-194 • (TG)			
Week 4	Day 151	Take **Chapter 19 Test** • Page 299 • (TG)			
	Day 152	Complete Chapter 20 Discussion Questions • Pages 195-196 • (TG) Chapter 20: Birds • Read pages 161-168 • (ETGE)			
	Day 153	Complete Chapter 20 Worksheet • Pages 197-198 • (TG)			
	Day 154	Take **Chapter 20 Test** • Page 301 • (TG)			
	Day 155	Complete Chapter 21 Discussion Questions • Pages 199-200 • (TG) Chapter 21: Mammals • Read pages 169-182 • (ETGE)			
Week 5	Day 156	Complete Chapter 21 Worksheet 1 • Pages 201-202 • (TG)			
	Day 157	Complete Chapter 21 Worksheet 2 • Pages 203-204 • (TG)			
	Day 158	Take **Chapter 21 Test** • Page 303 • (TG)			
	Day 159	Complete Chapter 22 Discussion Questions • Pages 205-206 • (TG) Chapter 22: Cone-Bearing Plants • Read pages 183-196 • (ETGE)			
	Day 160	Complete Chapter 22 Worksheet • Pages 207-208 • (TG)			
Week 6	Day 161	Take **Chapter 22 Test** • Page 305 • (TG)			
	Day 162	Complete Chapter 23 Discussion Questions • Pages 209-210 • (TG) Chapter 23: Spore-Forming Plants • Read pages 197-208 • (ETGE)			
	Day 163	Complete Chapter 23 Worksheet • Page 211 • (TG)			
	Day 164	Take **Chapter 23 Test** • Page 307 • (TG)			
	Day 165	Complete Chapter 24 Discussion Questions • Pages 213-214 • (TG) Chapter 24: Flowering Plants • Read pages 209-230 • (ETGE)			
Week 7	Day 166	Complete Chapter 24 Worksheet • Pages 215-216 • (TG)			
	Day 167	Take **Chapter 24 Test** • Page 309 • (TG)			
	Day 168	Complete Chapter 25 Discussion Questions • Pages 217-218 • (TG) Chapter 25: Coming Full Circle — My Conclusions Read pages 231-243 • (ETGE)			
	Day 169	Complete Chapter 25 Worksheet • Pages 219-220 • (TG)			
	Day 170	Take **Chapter 25 Test** • Page 311 • (TG)			
Week 8	Day 171	Study Day-Chapters 12-25			
	Day 172	Study Day-Chapters 12-25			
	Day 173	Study Day-Chapters 12-25			
	Day 174	Take **Sectional Exam 2** Chapters 12-25 • Pages 339-342 • (TG)			
	Day 175	Watch DVD in its entirety as a summary.			

Date	Day	Assignment	Due Date	✓	Grade
Week 9	Day 176	Study Day-Chapters 1-25			
	Day 177	Study Day-Chapters 1-25			
	Day 178	Study Day-Chapters 1-25			
	Day 179	Study Day-Chapters 1-25			
	Day 180	Take **Comprehensive Exam** Chapters 1-25 • Pages 343-348 • (TG)			
		Second Semester Grade			
		Final Grade			

Worksheets

for Use with

Evolution: The Grand Experiment

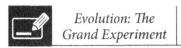

Purpose of Chapter: The purpose of Chapter 1 is to help students understand that there is more than one point of view regarding how all forms of life came about. It is important, in a diverse culture such as ours, to understand other points of view and to understand how others have arrived at their conclusions.

Answer the Discussion Questions. Compare your answers to the next page:

1. What are the two opposing views concerning how life came about and how humans came into being?

2. Do you think parents want creationism to be taught in *public* schools along with evolution? Why or why not?

Compare your answers:

1. What are the two opposing views concerning how life came about and how humans came into being?

   Answer: Generally, they are:

   (1) That life (in the form of a theoretical bacterium-like organism) came about spontaneously as a result of the big bang, eventually evolving into modern animals and humans over billions of years or

   (2) That all life was created by a higher power or deity not seen by human beings.

   (Note: A possible third view, believed by some, is that life evolved, but God helped the process along.)

2. Do you think parents want creationism to be taught in *public* schools along with evolution? Why or why not?

   Answer: According to the Gallup poll presented in this chapter, parents want creationism taught in public schools along with evolution so that students can learn the facts and evidences both for and against each theory.

1. Describe the three worldviews concerning how life and humans came into existence.

   a.

   b.

   c.

2. Define the big bang. (See Glossary in the student book.)

3. What year did Darwin publish his theory of evolution?

4. Describe the three major scientific developments concerning evolution that have occurred since Darwin first published his theory in 1859. Hint: Fossils, DNA, and genes.

   a.

   b.

   c.

5. Name the artist, the location, and the content of the famous artwork dealing with the origin of life, as seen on pages 2-3 of this chapter.

   a.

   b.

   c.

6. Since the middle of the 20th century there have been a growing number of ____________________ who reject the theory of evolution based on the discovery of processes and structures of which Darwin was unaware.

7. List the four best evidences against the theory of evolution cited by scientists who oppose evolution.

   a.

   b.

   c.

   d.

8. List the four best evidences for the theory of evolution cited by scientists who support evolution.

   a.

   b.

   c.

   d.

9. Describe the problems of teaching creationism and the theory of evolution to students using the dual model approach.

10. Describe the benefits of teaching the theory of evolution to students using the dual model approach.

11. Describe the results of the Gallup poll concerning the opinion of parents as to what should be taught in public schools. When parents were asked if creationism should be taught in *public* schools, what percentage said yes, what percentage said no, and what percentage were unsure?

    a.

    b.

    c.

    d.

Purpose of Chapter: The purpose of Chapter 2 is to show students that scientists are not infallible. Scientists can make mistakes just like anyone else.

Answer the Discussion Questions. Compare your answers to the next page:

1. Is it possible for a scientist to make a mistake? Give an example.

2. Have scientists made any mistakes in your lifetime?

3. If a majority of scientists believe in something, should you accept what they believe?

Compare your answers:

1. Is it possible for a scientist to make a mistake? Give an example.

   Answer: Yes. Scientists used to (incorrectly) believe the earth was the center of our solar system. (Galileo and others later showed that the sun was the center of our solar system.) Also, in the past, scientists thought that the earth was flat. We now know, of course, that the earth is round, like a ball.

2. Have scientists made any mistakes in your lifetime?

   Answer: Yes. Recent examples include cold fusion, the belief that certain foods cause certain diseases, etc.

3. If a majority of scientists believe in something, should you accept what they believe?

   Answer: No, not necessarily. Scientists can be wrong.

1. What theory did scientists believe *before* Darwin's theory of evolution was published? During what century was this theory first believed? How long was it perpetuated? In what year was it disproved?

   a.

   b.

   c.

   d.

2. Briefly describe the now disproved theory of spontaneous generation.

3. Describe Dr. von Helmont's "mice from dirty underwear" experiment that he offered as proof of the theory of spontaneous generation.

4. What would happen during von Helmont's time to anyone who dared to challenge spontaneous generation?

5. Describe the "maggots from rotting meat" experiment.

6. Describe Dr. Francesco Redi's experiment and tell whether his experiment was a proof for or against the theory of spontaneous generation.

   a.

   b.

7. Give the year of Dr. Francesco Redi's experiment.

8. Describe the "scum from clear pond water" experiment.

9. Describe the experiment of John Needham.

10. Describe Dr. Louis Pasteur's experiment and tell whether his experiment was a proof for or against the theory of spontaneous generation.

    a.

    b.

11. What year did Dr. Louis Pasteur disprove the spontaneous generation of bacteria from water?

12. What theory eventually replaced the theory of spontaneous generation and is the basis for the modern theory of how life began naturally?

13. What has spontaneous generation taught us, as a society, about the infallibility of scientists?

14. What has spontaneous generation taught us about scientific ideas?

Purpose of Chapter: The purpose of Chapter 3 is to teach students that one of Charles Darwin's proposed mechanisms for evolution — the law of use and disuse — was disproved in 1889, seven years after Charles Darwin died. Yet, this disproved mechanism for evolution is still believed by some today.

Answer the Discussion Questions. Compare your answers to the next page:

1. If a man has large muscles from lifting weights, will his children be born with large muscles? Why or why not?

2. Charles Darwin thought that if you exercised a horse every day, it would have big, strong muscles and the offspring of the horse would be born with stronger muscles. Was he right or wrong?

3. If stronger or larger muscles (as a result of exercising animals or weight lifting in humans) do not result in passing these traits on to the next generation, how can evolution occur then? How can improvements be passed on to the next generation?

Compare your answers:

1. If a man has large muscles from lifting weights, will his children be born with large muscles? Why or why not?

   Answer: No. No matter how large you build your muscles, your children are always born with the normal amount of muscle. The arm muscle cells are not passed on to the next generation, only the reproductive cells are. The DNA of the eggs or sperm of an animal are not affected by the muscle cells being exercised in the arms.

2. Charles Darwin thought that if you exercised a horse every day, it would have big, strong muscles and the offspring of the horse would be born with stronger muscles. Was he right or wrong?

   Answer: He was wrong. No matter how much you exercise a horse, its offspring are always born with normal-sized muscles. Darwin did not understand DNA and inheritance. He was wrong on this point.

3. If stronger or larger muscles (as a result of exercising animals or weight lifting in humans) do not result in passing these traits on to the next generation, how can evolution occur then? How can improvements be passed on to the next generation?

   Answer: Acquired characteristics was just one of the mechanisms that Darwin proposed to explain how evolution occurred. (This will be discussed in Chapter 4.)

1. What year did Darwin publish his first book on evolution?

2. Write out the complete title of Darwin's book, not just the shortened four-word title, *The Origin of Species*.

3. What organism did Darwin think evolved into all of the animals on the earth?

4. What organism, according to the modern theory of evolution, evolved into all of the animals on the earth?

5. List, in order, the six types of animals or organisms that theoretically evolved from one to another, from the theoretical primordial single-cell organism form to a *bird*, according to the modern theory of evolution.

   a. a single-cell organism

   b.

   c.

   d.

   e.

   f.

6. List, in order, the eight types of animals or organisms that theoretically evolved from one to another, from the theoretical primordial single-cell organism form to a *human being*, according to the modern theory of evolution.

   a. a single-cell organism

   b.

   c.

   d.

   e.

   f.

   g.

   h.

7. In general, how long did it take for a single-cell organism (or bacterium-like organism) to theoretically evolve into a human, according to the modern theory of evolution? How long did the theory of spontaneous generation propose it took to generate life, such as mice, maggots or bacteria?

   a.

   b.

8. Write out the definition of "acquired characteristics."

9. Give the other two names for "acquired characteristics."

   a.

   b.

10. Write out Charles Darwin's quote stating his belief in acquired characteristics and cite the book in which he wrote it.

    a.

    b.

11. Describe why characteristics acquired during a person's lifetime, such as large muscles or a sun tan, cannot be passed on to the next generation.

12. Describe four specific examples of acquired characteristics believed in the past and explain why each one was scientifically incorrect (pages 25–28).

    a.

    b.

    c.

    d.

13. Write Darwin's quote that shows his belief in acquired characteristics, the year he wrote it, and the source of the quotation.

    a.

    b.

    c.

14. Describe which famous scientist finally disproved acquired characteristics and how he did it.

15. In what year did August Weisman carry out his experiment?

Purpose of Chapter: The purpose of Chapter 4 is twofold: (1) To explain the concept of natural selection, the major mechanism for evolution proposed by Darwin, and the controversy surrounding it and (2) To explain to the students that the modern theory of evolution proposes that complex animals, such as whales or bats, evolved from land mammals as a result of pure chance/blind/undirected/random/accidental mutations.

Answer the Discussion Questions. Compare your answers to the next page:

1. Charles Darwin thought that nature could create totally different species of animals by killing off the weaker animals within that species. Do you think this is possible?

2. Charles Darwin believed that a black bear could have evolved into a whale. Do you think this is possible?

3. The theory of evolution teaches that one animal evolved into another animal by accidental mutations in the DNA. If you were exposed to a nuclear bomb's radioactivity or X-rays, both of which cause mutations in the DNA, would this increase the chance that your offspring would be more fit or healthy than you are now?

Compare your answers:

1. Charles Darwin thought that nature could create totally different species of animals by killing off the weaker animals within that species. Do you think this is possible?

   Answer: The student should try to answer this for themselves. The answer is in the lesson.

2. Charles Darwin believed that a black bear could have evolved into a whale. Do you think this is possible?

   Answer: The student should try to answer this for themselves. The answer is in the lesson.

3. The theory of evolution teaches that one animal evolved into another animal by accidental mutations in the DNA. If you were exposed to a nuclear bomb's radioactivity or X-rays, both of which cause mutations in the DNA, would this increase the chance that your offspring would be more fit or healthy than you are now?

   Answer: No. Exposure to radioactivity or X-rays increase the chance for accidental mutations (or damage to the DNA) resulting in an increased risk of diseases, such as leukemia and cancer, and a greater possibility of birth defects in your offspring.

1. What are the two different types of breeding?

   a.

   b.

2. How can artificial and natural breeding remove certain traits (color or size) from a breed but not add completely new body parts, such as a wing or a gill?

3. Describe how, under certain circumstances, nature might breed for a certain trait (e.g, fur color).

4. Explain the "limits of variability."

5. Name the scientist who discovered that new traits, such as albino eyes, could only come about by accidental mutations (in the DNA).

6. Give the year Dr. Morgan discovered accidental mutations and at what university his experiment took place.

   a.

   b.

7. Describe what Dr. Morgan observed that made him realize that accidental mutations could cause new traits.

8. Mutations can be dangerous to an animal's health. List four diseases in human beings that are the result of a single DNA letter mutation.

   a.

   b.

   c.

   d.

9. How many letters of DNA in a reproductive cell would have to be accidentally added or changed to bring about a new body system, such as a wing with feathers, or a cardiovascular system with a heart, blood vessels, and lungs?

10. Describe the controversy concerning accidental mutations causing new body systems. Specifically, describe why scientists who oppose evolution do not believe a series of random mutations could bring about a new animal or body system, such as a wing with feathers. Cite the blindfolded child example.

    a.

    b.

11. Explain why modern evolution scientists believe the term "adaptation" should be eliminated from the scientific vocabulary.

12. Explain why the term "adaptation" should always be replaced with the words "mutation and natural selection" or "fortuitous adaptation and natural selection" in order to be more accurate.

13. Cite three examples of modern scientists incorrectly using the concept of adaptation. (Do NOT memorize specific scientist names or exact quotes. Rather, memorize three general ideas mentioned in their quotes and WHY they are wrong.)

a.

b.

c.

14. Name the three groups of aquatic mammals which are members of the animal order "Cetacea."

a.

b.

c.

15. What is the size range (shortest to longest) of members of the order Cetacea?

16. Fill in the following information regarding the blue whale:

Length______________

Weight______________

Tongue size__________

Blood vessel size________

Size of heart____________

Is the blue whale the largest animal to have ever lived on earth?

17. Describe Charles Darwin's ideas concerning the land mammal from which whales could have evolved.

18. Name the scientist, living in England at the same time as Charles Darwin, who opposed Charles Darwin on his ideas about whales evolving from black bears.

19. Name three animals from which modern scientists have suggested whales evolved.

a.

b.

c.

20. Scientists now theorize that __________________ evolved from a land animal through a complicated series of chance mutations in the DNA of the reproductive cells.

21. Modern scientists who oppose evolution think the idea of a land animal becoming a whale by a series of accidental mutations in the DNA is even more __________________ than Darwin's idea that a bear could become a whale through acquired characteristics. They argue that the odds of a land animal changing into a whale, by a series of mistakes in the DNA, are statistically impossible.

22. Name the changes that would be required for a hyena to evolve into a whale.

    a.

    b.

    c.

    d.

    e.

    f.

    g.

    h.

    i.

23. Write the formula for calculating the number of letters of DNA that must change for one animal to theoretically evolve into another.

24. Name the four different letters that make up DNA.

    a.

    b.

    c.

    d.

25. Describe what the odds are for correctly changing a single letter of DNA. Two letters? Three letters?

    a.

    b.

    c.

26. Give the approximate number of DNA letters which would have to change, by chance, for a hyena-like animal to evolve into a whale.

27. Compare the odds of a hyena becoming a whale to the odds of winning the National Powerball Lottery 200 years in a row. Which is more likely?

    a.

    b.

Purpose of Chapter: The purpose of Chapter 5 is to show students that the concept of similarities between animals is used as one of the basic evidences for the theory of evolution. For instance, a chimpanzee and a human are similar in some of their body characteristics; because of these similarities, it is thought by some scientists who support evolution that humans evolved from apes. Scientists who oppose evolution suggest that similarities cannot be used as a proof for evolution because many unrelated animals (such as a shark and a dolphin, a seal and a sea lion, etc.) also appear similar.

Answer the Discussion Questions. Compare your answers to the next page:

1. In what ways do a chimpanzee and a human superficially look similar?

2. Does the fact that humans and chimps superficially look alike prove that evolution occurred, that humans evolved from apes?

3. Turn to page 56, the second page of Chapter 5, and look at the pictures of the shark and dolphin. A shark and a dolphin superficially look similar. Both have a dorsal fin, both have pectoral fins, and both have a widened tail for swimming.

   Evolution scientists do not believe they are closely related because a shark is a fish and a dolphin is a mammal. How can similarities between animals, such as an ape and a human, be used as proof *for* evolution when other animals, such as a shark and a dolphin, are thought not to be related yet also have similarities?

Compare your answers:

1. In what ways do a chimpanzee and a human superficially look similar?

   Answer: Answers may vary but should be thoughtful and based on observation.

2. Does the fact that humans and chimps superficially look alike prove that evolution occurred, that humans evolved from apes?

   Answer: Students should try to answer this for themselves. The answer is in the lesson.

3. Turn to page 56, the second page of Chapter 5, and look at the pictures of the shark and dolphin. A shark and a dolphin superficially look similar. Both have a dorsal fin, both have pectoral fins, and both have a widened tail for swimming.

   Evolution scientists do not believe they are closely related because a shark is a fish and a dolphin is a mammal. How can similarities between animals, such as an ape and a human, be used as proof *for* evolution when other animals, such as a shark and a dolphin, are thought not to be related yet also have similarities?

   Answer: Answers may vary but should be thoughtful and include examples.

1. Write out the definition of "related animals" as used by evolution scientists.

2. Write out the definition of "unrelated animals" as used by evolution scientists.

3. Describe why evolution scientists believe a chimpanzee, a dinosaur, a bird, and a whale are "related animals." Specifically, what is similar in these four animals?

4. Describe why evolution scientists believe a shark and dolphin are "unrelated animals" even though they appear similar, both having a dorsal fin, pectoral fins, and a widened tail for swimming.

5. Write out four similarities of the giant panda and the red panda (page 58).
   a.

   b.

   c.

   d.

6. Describe how evolution scientists came to believe the giant panda and the red panda are "unrelated animals" even though they have four similar features.

7. Write out the two similarities of the seal and the sea lion (page 59).

   a.

   b.

8. Name the two animals evolution scientists believe seals may have evolved from and the two animals evolution scientists believe sea lions may have evolved from. How does this make seals and sea lions unrelated animals?

   a. Seals-

   b. Sea Lions-

   c.

9. What animals have ears similar to a hyrax?

10. What animals have teeth similar to a hyrax?

11. Write out the definition of "convergent evolution."

12. Describe the three similarities shared by sharks, dolphins, and extinct aquatic reptiles (ichthyosaurs).

13. Explain why similarities in the dolphin, the shark and the ichthyosaur do not equate to evolution according to evolution scientists.

14. Name three animals with wings that evolution scientists believe are unrelated (pages 64–65).

a.

b.

c.

15. Name three animals with eyes that evolution scientists believe are unrelated.

a.

b.

c.

16. Name three animals with duckbills that evolution scientists believe are unrelated.

a.

b.

c.

17. Name three animals with eye rings that evolution scientists believe are unrelated.

a.

b.

c.

18. Name two animals with head crests that evolution scientists believe are unrelated.

    a.

    b.

19. Name two sets of animals that appear nearly identical yet evolution scientists believe are unrelated.

    a.

    b.

20. According to the theory of evolution, is the placental mole more closely related to a pouched mole or a whale?

21. According to the theory of evolution, is the placental mouse more closely related to a pouched mouse or a horse?

22. Explain why scientists who oppose evolution believe that similarities between two animals cannot be used as evidence for evolution.

Purpose of Chapter: Evolution scientists, including Charles Darwin, have attributed gaps in the fossil record to an inadequate collection of fossils. The purpose of Chapter 6 is to take a closer look at this line of reasoning. Specifically, this chapter addresses the question: Have enough fossils been collected by museums to judge or evaluate the theory of evolution?

Answer the Discussion Questions. Compare your answers to the next page:

Look at the numbers:

100

1,000

10,000

100,000

1,000,000

10,000,000

100,000,000

100,000,000+

Circle your answer:

1. How many fossils have been collected by museums worldwide?

   10,000? 100,000? 1,000,000? 10,000,000? 100,000,000? More?

2. How many fossil fish have been collected by museums so far?

   100? 1,000? 10,000? 100,000? 1,000,000? 10,000,000? 100,000,000? More?

3. How many fossil birds have been collected by museums so far?

   100? 1,000? 10,000? 100,000? 1,000,000? 10,000,000? 100,000,000? More?

4. How many fossil turtles have been collected by museums so far?

   100? 1,000? 10,000? 100,000? 1,000,000? 10,000,000? 100,000,000? More?

5. In your opinion, after learning about these numbers, have enough fossils been collected by museums to judge or evaluate the theory of evolution?

6. With 100,000,000[1] *collected* fossils, what should we see in the fossils, if evolution is true?

7. Using this same number of 100,000,000[2] *collected* fossils, what would you expect to see in the fossil record if evolution is, in fact, *not* true?

1 Older editions of *Evolution: The Grand Experiment* may have 200 million or "hundreds of millions" as the number of collected fossils. See footnote 1 in Appendix A of *Evolution: The Grand Experiment.*

2 ibid

Compare your answers:

1. 1,000,000,000[1] of the best/most representative fossils have been *collected* by museums so far.

2. 500,000 fossil fish have been *collected* by museums.

3. 200,000 fossil birds have been *collected* by museums.

4. 100,000 fossil turtles have been *collected* by museums.

5. In your opinion, after learning about these numbers, have enough fossils been collected by museums to judge or evaluate the theory of evolution?

   Answer: Answers may vary but should be thoughtful and include the reasons the student has this opinion.

6. With 1,000,000,000[2] *collected* fossils, what should we see in the fossils, if evolution is true?

   Answer: *This is an important point.* Here is the answer: If evolution occurred and if the fossil record is great, we should see one animal slowly changing into another over millions of years.

7. Using this same number of 1,000,000,000[3] *collected* fossils, what would you expect to see in the fossil record if evolution is, in fact, *not* true?

   Answer: *This is an important point.* Here is the answer: If evolution did not occur and if the fossil record is great, we would not see one animal changing into another. There would be ancestral gaps between animals.

---

1 Older editions of *Evolution: The Grand Experiment* may have 200 million or "hundreds of millions" as the number of collected fossils. See footnote 1 in Appendix A of *Evolution: The Grand Experiment.*

2 ibid

3 ibid

1. Define the term "fossil record."

2. What was Charles Darwin's dilemma about the fossil record when he wrote *The Origin of Species*?

3. Write out Charles Darwin's quote acknowledging the discrepancy in the fossil record.

4. What argument did Charles Darwin make in the first two chapters of *The Origin of Species* about the fossil record?

5. Describe what the fossil record would show if the fossil record was nearly complete and if evolution was true.

6. Describe what the fossil record would show if the fossil record was nearly complete and if evolution was not true.

7. How many fossils (total number of all types of fossils) have been *collected* by museums since Darwin's time?

8. Fill in the number of fossils found in this list. **Memorize this list.**
   a. The number of fossil fish in museums:
   b. The number of fossil dinosaurs in museums:
   c. The number of fossil bats in museums:
   d. The number of fossil bird specimens in museums:
   e. The number of fossil pterosaurs (flying reptiles) in museums:
   f. The number of fossil insects in museums:
   g. The number of fossil plants in museums:
   h. The number of fossil turtles in museums:
   i. The number of fossil whales:
   j. The number of fossilized soft-bodied animals, plants, and other soft-bodied organisms that have been found:

9. Give examples of fossilized soft-bodied plant parts or animals that have been collected by museums.

10. How many invertebrates have been *collected* by museums? How many vertebrates?

    a.

    b.

11. What are scientists saying, in general, about the fossil record today?

12. What is the percentage of *orders* of land vertebrates living today that have also been found as fossils?

13. What is the percentage of *families* of land vertebrates living today, *including* birds, that have also been found as fossils?

14. What is the percentage of *families* of land vertebrates living today, *excluding* birds, that have also been found as fossils?

15. Write a six-sentence essay answer to these questions: In your opinion, is the fossil record today adequate to judge the two models detailed on pages 75–76? Have enough fossils been *collected* now to make this judgment? Why or why not?

Purpose of Chapter: To describe the sudden appearance of invertebrates in the fossil record and compare this to Darwin's predictions.

Answer the Discussion Questions. Compare your answers to the next page:

1. What is a vertebrate?

2. What is an invertebrate?

3. Give an example of an invertebrate.

4. How many fossilized invertebrates have been *collected* by museums so far?

5. If evolution is true, and with this many fossils in museums, would you expect to see the evolution of invertebrates in the fossils (i.e., one invertebrate slowly changing into another)? Why or why not?

6. If evolution is not true, and if the fossil record is very good, what should be found in the fossil record regarding the origin of invertebrates?

7. If the fossil record is very good, and if the theory of evolution is true, what should be found regarding the evolution of a lobster from a bacterium? This is a review of the charts on pages 75–76 of Chapter 6.

8. If the fossil record is very good, and if the theory of evolution is not true, what should be found regarding the evidence of evolution of a lobster from a bacterium? This is a review of the charts on pages 75–76 of Chapter 6.

Compare your answers:

1. What is a vertebrate?

   Answer: Animal with a backbone.

2. What is an invertebrate?

   Answer: Animal without a backbone or spinal cord.

3. Give an example of an invertebrate.

   Answer: Jellyfish, insects, shellfish, shrimp, lobster, crabs, bacteria, worms, etc.

4. How many fossilized invertebrates have been collected by museums so far?

   Answer: Scientists have collected over 750,000,000 fossil invertebrates including millions of fossilized microscopic organisms, such as fossil bacteria.

5. If evolution is true, and with this many fossils in museums, would you expect to see the evolution of invertebrates in the fossils (i.e., one invertebrate slowly changing into another)? Why or why not?

   Answer: Answers will vary but should be thoughtful and present logical reasoning.

6. If evolution is not true, and if the fossil record is very good, what should be found in the fossil record regarding the origin of invertebrates?

   Answer: Invertebrates would appear suddenly in the fossil record with no intermediate ancestors. The fossil record would not show the transitional forms between a bacterium and a lobster. There would be ancestral gaps between the two.

7. If the fossil record is very good, and if the theory of evolution is true, what should be found regarding the evolution of a lobster from a bacterium? This is a review of the charts on pages 75–76 of Chapter 6.

   Answer: We would find a bacterium slowly changing into a lobster over time, with lots of intermediate or transitional animals in between, such as (1) an animal one quarter of the way between a bacterium and a lobster, (2) an animal halfway between a bacterium and a lobster, and (3) an animal three quarters of the way between a bacterium and a lobster. For example, if evolution is true, we should see the slow appearance of the lobster's shell over time, the slow appearance of the lobster's claws over time, the slow appearance of the lobster's legs over time, the slow development of the lobster's eyes over time, etc.

8. If the fossil record is very good, and if the theory of evolution is *not* true, what should be found regarding the evidence of evolution of a lobster from a bacterium? This is a review of the charts on pages 75–76 of Chapter 6.

   Answer: We would find a bacterium and we would find a fossil lobster, but we would not find fossil animals in between.

1. Define "invertebrate."

2. Define "vertebrate" (Found in Glossary).

3. How many trilobite fossils have been collected by museums?

4. How many ancestors of trilobites have been discovered so far?

5. Name a soft-bodied invertebrate soft coral fossil that has been discovered in the Ediacaron layer that lives today.

6. Explain why trilobites are problematic for the theory of evolution.

7. How many fossil invertebrates have been collected by museums?

8. What is the Cambrian Explosion?

9. Study the fossil record of invertebrates. Does the fossil record match Darwin's predictions?

10. Provide the explanation, given by scientists who support evolution, for why there is a lack of ancestors for invertebrates.

11. Explain why scientists who oppose evolution believe the fossil record of invertebrates suggests that evolution did not occur.

12. Name the five lowest fossil layers starting with the Ediacaran fossil layer as the lowest, according to scientists who support evolution.

    a.

    b.

    c.

    d.

    e.

13. How many fossil sea pens have been found in the Ediacaran fossil layer?

14. How many ancestors of fossil sea pens have been found?

15. Study the fossil record in Chapters 6 and 7 and decide if the fossil record of fossil sea pens, jellyfish, and sponges matches the predictions of the theory of evolution or not.

Purpose of Chapter: To describe the fossil record of fish (the theoretical first vertebrates) and compare the fossil record to Darwin's prediction. Darwin predicted that as more fossil fish and invertebrates were found, the evolutionary intermediate ancestors of fish would also be found. This means that the fossil record would eventually show an invertebrate, such as a jellyfish or starfish, slowly changing into a fish. According to both evolution scientists and scientists who oppose evolution, the predicted evolutionary ancestors of fish (vertebrates) are missing.

Answer the Discussion Questions. Compare your answers to the next page:

1. What is a vertebrate? Give several examples.

2. Is a fish a vertebrate or an invertebrate?

3. Describe the differences between vertebrates and invertebrates.

4. If evolution is true, and if the fossil record is abundant and representative of what happened in the past, what should you see in the fossil record between invertebrates and vertebrates?

5. Over 100,000,000 fossil invertebrates have been collected by museums and 500,000 fossil fish have been collected by museums. With this many fossil invertebrates and this many fossil fish (vertebrates), would you expect to see the evidence of an invertebrate evolving into a fish?

6. Have you ever seen (at a museum or in a book) a display showing the evolution of an invertebrate into a fish? If so, at what museum in what city?

7. If no one has seen this, what does this mean about fish evolution? Did it occur?

Compare your answers:

1. What is a vertebrate? Give several examples.

   Answer: A vertebrate is an animal with a backbone. The individual bones of the backbone are called vertebrae. Examples of vertebrates include humans, fish, frogs, snakes, apes, dogs, and cats.

2. Is a fish a vertebrate or an invertebrate?

   Answer: All fish are vertebrates since they have backbones.

3. Describe the differences between vertebrates and invertebrates.

   Answer: Vertebrate animals have brains and spinal cords. The brains are surrounded by skulls and the spinal cord is surrounded by vertebrae. Invertebrates do not have a spinal cord or vertebrae.

4. If evolution is true, and if the fossil record is abundant and representative of what happened in the past, what should you see in the fossil record between invertebrates and vertebrates?

   Answer: You should see an invertebrate animal, such as jellyfish or starfish, slowly forming a spinal cord, slowly forming a brain, slowly forming a skull, and slowly forming vertebrae.

5. Over 100,000,000 fossil invertebrates have been collected by museums and 500,000 fossil fish have been collected by museums. With this many fossil invertebrates and this many fossil fish (vertebrates), would you expect to see the evidence of an invertebrate evolving into a fish?

   Answer: Answers will vary but should be thoughtful and present logical reasoning.

6. Have you ever seen (at a museum or in a book) a display showing the evolution of an invertebrate into a fish? If so, at what museum in what city?

   Answer: No such museum exhibit exists according to expert Dr. John Long.

7. If no one has seen this, what does this mean about fish evolution? Did it occur?

   Answer: Answers will vary but should be thoughtful and present logical reasoning.

1. How many fossilized fish have been *collected* by museums?
2. How much detail can be seen in fish fossils?
3. How many invertebrate fossils have been *collected* by museums?
4. Discuss the following points regarding the fish evolution chart on page 96:
   a. Has the common ancestor of fish been discovered?
   b. Have the intermediate stages between the theoretical common ancestor of all fish and the different fish families been found?
   c. Does the fossil record of fish show fossil evidence of one fish changing into another fish or do fish families appear suddenly without fossilized transitional ancestors?
5. How many intermediate animals have been found showing the intermediate stages between an invertebrate and a fish (or vertebrate)?
6. Describe what is being said by scientists who *oppose* evolution about the proposed evolutionary transitional forms for each fish group.
7. Dr. John Long, paleontologist and Head of Science at the Museum Victoria, Melbourne, Australia, is a strong proponent of fish evolution and author of the book *The Rise of Fishes: 500 Million Years of Evolution*. What did he say about the fossil evidence for the transition from invertebrates to the first backboned fish?
8. Summarize Dr. John Long's explanation of the fossil evidence for the origin of the first jawed fish.

9. Summarize Dr. John Long's explanation of the fossil evidence for the origin of sharks.

10. Summarize Dr. John Long's explanation of the fossil evidence for the origin of spiny fin fish.

11. Summarize Dr. John Long's explanation of the fossil evidence for the origin of bony fish.

12. Summarize Dr. John Long's explanation of the fossil evidence for the theoretical evolutionary interrelationships of the major groups of fish.

13. Describe which model the fossil record best matches.

Purpose of Chapter: To describe the fossil record of bats and compare this fossil record to Darwin's prediction. Darwin predicted that as more fossils were found, the evolutionary intermediate ancestors for bats would also be found. This means that the fossil record would eventually show a ground animal (possibly about the size of a mouse) slowly changing into a bat. According to both evolution scientists and scientists who oppose evolution, all of the predicted evolutionary ancestors of bats are missing. (See pages 75–76 of Chapter 6 for a further discussion of the predictions of the fossil record.)

Answer the Discussion Questions. Compare your answers to the next page:

1. What kind of animal is a bat? Is it a bird? A reptile? A mammal? Why?

2. How do bats catch flying insects at night when it is completely pitch black?

3. The theory of evolution suggests that bats (flying mammals) arose from a non-flying mammal, a mammal possibly about the size of a mouse. Describe what would have to occur for a land mammal, say a mouse, to become a bat. What changes would have to occur? List five of these changes.

4. Could an animal, such as a mouse, develop wings simply by repeatedly moving its front arms up and down? Why or why not?

5. Could an animal, such as a mouse, develop new structures (such as wings), using the principle of the survival of the fittest, also known as natural selection?

6. If a mouse-like animal could not evolve into a bat by jumping off small ledges or by natural selection, then how does the theory of evolution suggest an animal similar to a mouse could develop wings and change into a bat?

7. Do you think an animal could develop wings, sonar-like echolocation, hollow bones, and so forth, by chance mutations in the DNA of the reproductive cells, as the theory of evolution suggests?

Compare your answers:

1. What kind of animal is a bat? Is it a bird? A reptile? A mammal? Why?

   Answer: Bats are mammals because they are warm-blooded, have hair, and suckle their young, as do all mammals.

2. How do bats catch flying insects at night when it is completely pitch black?

   Answer: Bats have a sophisticated sonar-like system called echolocation.

3. The theory of evolution suggests that bats (flying mammals) arose from a non-flying mammal, a mammal possibly about the size of a mouse. Describe what would have to occur for a land mammal, say a mouse, to become a bat. What changes would have to occur? List five of these changes.

   Answer: The ground animal would have to form wings. It would have to grow extra long fingers to form the wings. It would have to form a cape-like membrane on each arm. Its bones would have to change from solid bones for walking on the ground into lightweight hollow bones for flight. It would have to transfer its largest muscles from the back legs for running to the front legs to power the wings. It would have to develop a sonar-like echolocation system for catching insects in flight.

4. Could an animal, such as a mouse, develop wings simply by repeatedly moving its front arms up and down? Why or why not?

   Answer: No. The law of acquired characteristics (the law of use), which was believed by Darwin, was disproved in 1889. (See page 30.)

5. Could an animal, such as a mouse, develop new structures (such as wings), using the principle of the survival of the fittest, also known as natural selection?

   Answer: No. Natural selection only removes certain traits. It does not cause the development of completely new body structures, such as wings or sonar-like echolocation.

6. If a mouse-like animal could not evolve into a bat by jumping off small ledges or by natural selection, then how does the theory of evolution suggest an animal similar to a mouse could develop wings and change into a bat?

   Answer: The theory of evolution says that a ground mammal changed into a bat by a series of mistaken mutations in the DNA of the reproductive cells. For this to occur, thousands of letters of DNA would have had to change by accident, in the proper location, and in the proper order.

7. Do you think an animal could develop wings, sonar-like echolocation, hollow bones, and so forth, by chance mutations in the DNA of the reproductive cells, as the theory of evolution suggests?

   Answer: Answers will vary but should be thoughtful and present logical reasoning.

1. How many fossilized bats have scientists found so far?

2. What should be found in the fossil record of bats if evolution is true? Give a description of what the theoretical intermediate animals would have looked like.

3. Predict what should be found in the fossil record of bats if evolution is not true.

4. How many of the theoretical evolutionary ancestors of bats have been found?

5. Has the ground mammal from which a bat theoretically evolved been found?

6. Describe why no theoretical evolutionary ancestors of bats have been found, according to scientists who *support* evolution.

7. Explain why no theoretical evolutionary ancestors of bats have been found, according to scientists who *oppose* evolution.

8. Give the name of the fossil layer in which bats first appear. (The age and names of fossil layers are controversial and not agreed on by all scientists.)

9. Compare the appearance of the oldest fossil bats found (to date) to the appearance of modern bats.

10. Were any of the fossil bats discovered so far the intermediate ancestors predicted by Darwin? In other words, were any of the bats discovered non-flying or non-functional bats?

11. Describe how, according to Dr. Habersetzer, evolution scientists have determined when bat evolution started and what happened in the process of bat evolution.

Author's Note: After the first edition of this book was published in the fall of 2007, a new fossil bat, *Onychonycteris finneyi*, was reported in the February 2008 edition of the journal *Nature*. Turn now to Appendix D: Bat Evolution Update for a detailed discussion of this important fossil bat.

Purpose of Chapter: To study the fossil record of seals and sea lions in relation to the theory of evolution. Darwin predicted that as more fossils were found, the evolutionary intermediate ancestors for seals and sea lions would also be found. This means that the fossil record would eventually show a land mammal slowly changing into a seal or sea lion. According to evolution scientists and scientists who oppose evolution, the predicted evolutionary ancestors of seals and sea lions are missing.

Answer the Discussion Questions. Compare your answers to the next page:

1. How deep do you think a submarine and a seal can dive? Which can dive deeper in the ocean, a seal or a nuclear submarine?

2. What kind of animals are seals and sea lions? Are they fish, mammals, or something else? Why?

3. What land mammal do you think evolution scientists today believe evolved into a sea lion?

4. If the theory of evolution were true, what changes, by mutations, would have to occur for a dog-like animal to evolve into a sea lion? List four changes.

   a.

   b.

   c.

   d.

5. How many fossil sea lions have been found?

6. With such a rich fossil record of sea lions, what would you expect to see in the fossil record if evolution is true?

7. With such a rich fossil record of sea lions, what would you expect to see in the fossil record if evolution is *not* true?

Compare your answers:

1. How deep do you think a submarine and a seal can dive? Which can dive deeper in the ocean, a seal or a nuclear submarine?

   Answer: A seal can dive to 5,200 feet, but a nuclear submarine can only dive to 800 feet.

2. What kind of animals are seals and sea lions? Are they fish, mammals, or something else? Why?

   Answer: Seals and sea lions are mammals because they are warm-blooded, have some body hair on their face, and suckle their young, as do all mammals.

3. What land mammal do you think evolution scientists today believe evolved into a sea lion?

   Answer: Modern evolution scientists believe that a bear or dog-like animal evolved into a sea lion. Do you believe this?

4. If the theory of evolution were true, what changes, by mutations, would have to occur for a dog-like animal to evolve into a sea lion? List four changes.

   Answer: Some examples of changes that would need to occur for a dog-like animal to change into a sea lion: (a) its front legs would have to turn into finned flippers, by chance; (b) its back legs would have to turn into finned feet, by chance; (c) it would have to go from having lots of hair to becoming nearly hairless, by chance; and (d) it would have to develop the ability to hold its breath for long periods of time and develop the capacity to dive in very deep water, by chance.

5. How many fossil sea lions have been found?

   Answer: Thousands.

6. With such a rich fossil record of sea lions, what would you expect to see in the fossil record if evolution is true?

   Answer: You would expect to see the evolutionary intermediate ancestors between a land mammal (such as a bear-like or a dog-like animal) and a sea lion in the fossil record. You would expect to see fossils of land mammals slowly developing front fins, finned feet, and so forth.

7. With such a rich fossil record of sea lions, what would you expect to see in the fossil record if evolution is *not* true?

   Answer: You would expect to find only fossils of the dog-like or bear-like animal and fossils of sea lions, but no transitional forms between the two. There would be no fossils of animals showing the slow development over time of front fins, finned feet, etc.

Sea Lions

1. Distinguish a sea lion from a seal.

2. List at least three varieties of sea lions. How fast they can swim?

   a.

   b.

   c.

   d.

3. What two land animals (or mammals) do evolution scientists believe may have evolved into a sea lion?

   a.

   b.

4. Has the proposed land animal that evolved into a sea lion been found?

5. How many of the fossil intermediate (transitional) animals along the proposed evolutionary line from a land mammal to sea lions have been found?

6. How do evolution scientists respond to the absence of transitional fossils for sea lions?

7. Give the name of the earliest sea lion and describe what it looked like (page 109).

   a.

   b.

8. How many fossil sea lions have been found?

9. Using the chart below, describe the fossil record of the proposed evolution of sea lions.

   Let's assume Animal A is the theoretical land mammal which evolved into a sea lion. (See example on page 76 of Chapter 6.) On the blank lines to the right, place the number of evolutionary intermediate fossils found between Animals A through Animal H. If no intermediate ancestors have been found, place the number 0. Lastly, for Animal I, place the number of fossil sea lions found (page 110).

   (The first and last Animals have been written in to help you understand what is expected.)

Animal	The number of fossils found
Animal A (Land mammal that theoretically evolved into sea lion)	0
Animal B	
Animal C	
Animal D	
Animal E	
Animal F	
Animal G	
Animal H	
Animal I (Sea lion)	Thousands

Seals

10. a. How many feet deep can seals dive?

    b. Compare this to a nuclear submarine.

    c. Compare this to sea lions.

11. How long can seals hold their breath compared to a sea lion?

12. Give the two possible land animals (or mammals) evolution scientists believe seals may have evolved from (page 111).

    a.

    b.

13. Have the proposed land animals that may have evolved into seals been found?

14. How many of the fossil intermediate (transitional) animals along the proposed evolutionary line from a land mammal to seals have been found?

15. How many fossil seals have been found?

16. Does the fossil record of seals match Darwin's predictions? To answer this question, compare the fossil record of seals to the two models listed on pages 75–76 in Chapter 6.

17. Using the chart below, describe the fossil record of the proposed evolution of seals.

Let's assume Animal A is the theoretical land mammal which evolved into a seal. (See example on page 76 of Chapter 6.) On the blank lines to the right, place the number of evolutionary intermediate fossils found between Animals A through Animal H. If no intermediate ancestors have been found, place the number 0. Lastly, for Animal I, place the number of fossil seals found.

Animal	The number of fossils found
Animal A (Land mammal that theoretically evolved into seal)	0
Animal B	
Animal C	
Animal D	
Animal E	
Animal F	
Animal G	
Animal H	
Animal I (Number of fossil seals)	

Author's Note: Two years after the first edition of this book was published in the fall of 2007, a new fossilized "walking seal," named *Puijila darwini*, was reported in the April 2009 edition of the journal *Nature*. It was heralded as the missing link. Turn now to Appendix E: Pinniped Evolution Update for a more detailed discussion of this most important fossil!

Purpose of Chapter: To study the fossil record of pterosaurs (flying reptiles) in relation to the theory of evolution. Darwin predicted that as more fossils were found, the evolutionary intermediate ancestors for pterosaurs (flying reptiles) would also be found. The modern theory of evolution suggests that the fossil record would eventually show a land reptile slowly changing into a flying reptile. According to evolution scientists and scientists who oppose evolution, all the predicted evolutionary ancestors of pterosaurs (flying reptiles) are missing.

Answer the Discussion Questions. Compare your answers to the next page:

1. Have you ever seen a flying reptile (pterosaur) in a movie? If so, which movie?

2. What kind of animal is a pterosaur: a bird, a mammal, or a reptile?

3. Are pterosaurs alive today or are they extinct?

4. Does the fact that an animal is extinct prove evolution?

5. If evolution is true, and if the fossil record were nearly complete, what would you expect to see regarding the origin of pterosaurs from a ground reptile such as a lizard?

6. If evolution is not true, and if the fossil record were nearly complete, what would you expect to see in the fossils regarding the origin of pterosaurs?

7. What do you think the fossil record actually shows regarding the theoretical evolution of pterosaurs? Does the fossil record show evolution of pterosaurs or not?

Compare your answers:

1. Have you ever seen a flying reptile (pterosaur) in a movie? If so, which movie?

   Answer: Pterosaurs are in Jurassic Park III and King Kong. Others may apply.

2. What kind of animal is a pterosaur: a bird, a mammal, or a reptile?

   Answer: A flying reptile.

3. Are pterosaurs alive today or are they extinct?

   Answer: Pterosaurs are extinct. They lived during the time of the dinosaurs.

4. Does the fact that an animal is extinct prove evolution?

   Answer: The extinction of an animal, such as the pterosaurs or the dinosaurs, does not mean that evolution has, in fact, occurred. Extinction simply means an animal no longer exists.

5. If evolution is true, and if the fossil record were nearly complete, what would you expect to see regarding the origin of pterosaurs from a ground reptile such as a lizard?

   Answer: You would expect to see one reptile (a land reptile perhaps similar to a lizard) slowly changing into a flying reptile over millions of years of fossils. You would expect to see (1) the front legs of a lizard-like reptile slowly changing into wings, (2) the wing membrane slowly forming over millions of years, and (3) the solid bones of the ground lizard slowly becoming hollow in order to make the animal lighter for flight.

6. If evolution is not true, and if the fossil record were nearly complete, what would you expect to see in the fossils regarding the origin of pterosaurs?

   Answer: You would expect to see a large ancestral gap between the theoretical land reptile (a lizard-like animal) and pterosaurs. In other words, there would not be any fossil evidence of a lizard-like animal slowly changing into a pterosaur.

7. What do you think the fossil record actually shows regarding the theoretical evolution of pterosaurs? Does the fossil record show evolution of pterosaurs or not?

   Answer: Answers may vary but should give reasons for the answers given. (Answers are in Chapter 11.)

1. What kind of animal are pterosaurs? A bird? A mammal? A reptile?

2. What well-known land reptile lived at the same time as pterosaurs?

3. How many fossil pterosaurs have been found so far and on what continents have they been found?

4. Describe how well preserved some of the fossil pterosaurs are.

5. Name the two varieties of pterosaurs. How are they distinguished from each other?

   a.

   b.

6. Describe the range in size of flying reptiles, from the smallest pterosaur to the largest. Compare the size of the largest pterosaur to the size of a fighter jet.

   a.

   b.

7. Name the land animal scientists believe evolved into a pterosaur.

8. How many pterosaur direct ancestor fossils have been found?

9. How many fossil intermediate (transitional) animals have been found between a land reptile and a flying pterosaur?

10. Using chart below, describe the fossil record of pterosaurs.

    Let's assume Animal A is the theoretical land reptile which evolved into a pterosaur. (See example on page 76 of Chapter 6.) On the blank lines on the right, place the *number* of evolutionary intermediate fossils found for Animals A through Animal H. If no intermediate ancestors have been found, place the number 0. Lastly, for Animal I, place the number of fossil pterosaurs found.

Animal	The number of fossils found
Animal A (Land reptile which theoretically evolved into pterosaur)	0
Animal B	
Animal C	
Animal D	
Animal E	
Animal F	
Animal G	
Animal H	
Animal I (Flying reptile — Pterosaurs)	

11. Compare the answer to Question 10 to the predictions of the theory of evolution. (Pages 75–76 of Chapter 6.) Do the numbers of intermediate animals match the theory of evolution?

Purpose of Chapter: To study the fossil record of dinosaurs in relation to the theory of evolution and Darwin's predictions. Darwin predicted that as more fossils of dinosaurs were found, the evolutionary intermediate ancestors for dinosaurs would also be found. This means that the fossil record would eventually show a non-dinosaurian reptile, possibly an animal similar to an alligator, slowly changing into a dinosaur; and then later, over time, one dinosaur (family) evolving into another dinosaur (family). According to evolution scientists and scientists who oppose evolution, the predicted evolutionary ancestors of dinosaurs are missing.

Answer the Discussion Questions. Compare your answers to the next page:

1. What kind of dinosaur is "Sue," the famous dinosaur at the Chicago Field Museum?

2. How many *T. rex* dinosaurs do you think have been found and are now collected in museums around the world?

3. Do you think this is a good fossil record of *T. rex*?

4. Were dinosaurs reptiles? Mammals? Amphibians?

5. Assuming evolution is true, if *T. rex* evolved from a non-dinosaurian reptile, such as an animal similar to an alligator, and if the fossil record were nearly complete, what kind of fossils would you expect to find between the alligator-like reptile and *T. rex*?

6. How many of the theoretical evolutionary ancestors for *T. rex* have been found?

Compare your answers:

1. What kind of dinosaur is "Sue," the famous dinosaur at the Chicago Field Museum?

   Answer: Sue is a Tyrannosaurus rex.

2. How many *T. rex* dinosaurs do you think have been found and are now collected in museums around the world?

   Answer: There are 32 skeletons, 12 of which are nearly complete.

3. Do you think this is a good fossil record of *T. rex*?

   Answer: Generally, to have 32 copies of the same organism is considered a very good fossil record.

4. Were dinosaurs reptiles? Mammals? Amphibians?

   Answer: All dinosaurs were reptiles. They laid eggs and had scaly skin.

5. Assuming evolution is true, if *T. rex* evolved from a non-dinosaurian reptile, such as an animal similar to an alligator, and if the fossil record were nearly complete, what kind of fossils would you expect to find between the alligator-like reptile and *T. rex*?

   Answer: (See pages 75–76 in Chapter 6 for a further discussion of the predictions of the fossil record.) You would expect to find not only the fossil non-dinosaurian alligator-like reptile and the fossilized *T. rex*, but also many intermediate animals between the two. You would expect to find the ancestral fossilized animals getting bigger and bigger, from the size of a 12-foot-long alligator-like reptile to a 42-foot-long, 18,000-pound *T. rex*. The fossils would also show the transition from a reptile, walking on all four legs, to a *T. rex* dinosaur, walking on just its back legs.

6. How many of the theoretical evolutionary ancestors for *T. rex* have been found?

   Answer: Answers may vary but should be thoughtful and express logical reasoning.

1. Which dinosaur was the largest meat-eating dinosaur to have ever lived?

2. How many *T. rex* skeletons have been found?

3. How many of the theoretical intermediate evolutionary ancestors of *T. rex* have been found?

4. Using the chart below, describe the fossil record of the proposed evolution of *T. rex*.

   Let's assume Animal A is the first theoretical dinosaur to have ever lived which theoretically evolved into a *T. rex* dinosaur. (See example on page 76 of Chapter 6.) On the blank lines to the right, place the number of evolutionary intermediate fossils found between Animals A through Animal H. If no intermediate ancestors have been found, place the number 0. Lastly, for Animal I, place the number of fossil *T. rex* dinosaurs found.

Animal	The number of fossils found
Animal A (First dinosaur which theoretically evolved into *T. rex*)	0
Animal B	
Animal C	
Animal D	
Animal E	
Animal F	
Animal G	
Animal H	
Animal I (*T. rex*)	

5. Compare the answer to question 4 to the predictions of the theory of evolution. Do the actual fossil numbers match the theory of evolution?

6. What do evolution scientists mean when they call a dinosaur a "cousin" of another dinosaur?

7. List the names of the three fossil layers containing dinosaur bones, according to scientists who support evolution.

   a.

   b.

   c.

8. Which dinosaur was the largest horned dinosaur to have ever lived?

9. Name the two dinosaur groups based on the shape of the pelvis.

   a.

   b.

10. What type of dinosaur is *Triceratops* (based on the shape of its pelvis)?

11. How many *Triceratops* dinosaurs have been found?

12. How many of the proposed evolutionary ancestors of *Triceratops* have been found?

13. Using the chart below, describe the fossil record of the proposed evolution of *Triceratops* dinosaurs.

    Let's assume Animal A is the first theoretical dinosaur to have ever lived which theoretically evolved into a *Triceratops* dinosaur. (See example on page 76 of Chapter 6.) On the blank lines to the right, place the number of evolutionary intermediate fossils found between Animals A through Animal H. If no intermediate ancestors have been found, place the number 0. Lastly, for Animal I, place the number of fossil *Triceratops* dinosaurs found.

Animal	**The number of fossils found**
Animal A (First dinosaur which theoretically evolved into *Triceratops*)	0
Animal B	
Animal C	
Animal D	
Animal E	
Animal F	
Animal G	
Animal H	
Animal I *(Triceratops)*	

14. Compare the answer to question 13 to the predictions of the theory of evolution. Do the actual fossil numbers match the theory of evolution?

15. What was the former name of *Apatosaurus*?

16. What is the general name for dinosaurs with long necks?

17. How many *Apatosaurus* skeletons have been found?

18. How many of the proposed evolutionary ancestors of *Apatosaurus* have been found?

19. Using the chart below, describe the fossil record of the proposed evolution of *Apatosaurus*.

Let's assume Animal A is the first theoretical dinosaur to have ever lived and which theoretically evolved into an *Apatosaurus* dinosaur. (See example on page 76 of Chapter 6.) On the blank lines to the right, place the number of evolutionary intermediate fossils found between Animals A through Animal H. If no intermediate ancestors have been found, place the number 0. Lastly, for Animal I, place the number of fossil *Apatosaurus* dinosaurs found.

Animal	The number of fossils found
Animal A (First dinosaur which theoretically evolved into *Apatosaurus*)	0
Animal B	
Animal C	
Animal D	
Animal E	
Animal F	
Animal G	
Animal H	
Animal I *(Apatosaurus)*	

20. Compare the answer to question 19 to the predictions of the theory of evolution. Do the actual fossil numbers match the theory of evolution?

21. How many fossil dinosaur bones have been found?

22. How many individual fossil dinosaurs have been collected by museums?

23. How many dinosaur species have been found?

24. Has the theoretical evolutionary common ancestor of all dinosaurs been found?

25. How many species of dinosaurs have direct ancestors which have been found?

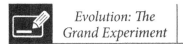

Purpose of Chapter: To see one of the "three best" fossil evidences for the theory of evolution — the evolution of whales from a land mammal — and make a critical evaluation of the evidence.

This chapter is unique in that the author of the book happened upon an important discovery during one particular interview on the subject of whales. Most scientists were unaware of this information, including many of the other whale experts interviewed for this chapter. Since the first edition of this book was released, museums have changed their whale evolution displays to correct the serious problems revealed in this chapter. The reader is encouraged to first read this chapter then immediately read Appendix F: Whale Evolution Update for the latest information on whale evolution.

Answer the Discussion Questions. Compare your answers to the next page:

1. What kind of animal is a whale? A fish or a mammal? Why?

2. Which land mammal did Charles Darwin suggest could have evolved into whales?

3. What land mammal do evolution scientists today believe evolved into a whale?

4. How strong is the evidence that a land mammal evolved into a whale?

Compare your answers:

1. What kind of animal is a whale? A fish or a mammal? Why?

   Answer: Whales are mammals because they suckle their young, they are warm-blooded, and they have hair (some facial hair used for tactile sensation).

2. Which land mammal did Charles Darwin suggest could have evolved into whales?

   Answer: Darwin conjectured that black bears could have evolved into whales.

3. What land mammal do evolution scientists today believe evolved into a whale?

   Answer: Answers may vary but students should include reasons for their answer. (The answer is in Chapter 13.)

4. How strong is the evidence that a land mammal evolved into a whale?

   Answer: Answers may vary but should be thoughtful and present logical reasoning.

1. What kind of animal is a whale? Is it a fish? A mammal? Why are whales classified as this type of animal?

2. Name the three types of Cetaceans.

   a.

   b.

   c.

3. According to the theory of evolution, what animal (type) evolved into land mammals?

4. From what kind of mammal did whales theoretically evolve — an aquatic mammal or a land mammal? When did this whale evolution theoretically happen?

5. Compare what evolution scientists are saying about the fossil evidence for the evolution of whales to the fossil evidence for the evolution of trilobites, fish, bats, and pterosaurs. (Page 88 for trilobites, page 96 for fish, page 100 for bats, and page 116 for pterosaurs.) Concerning the evidence for evolution, what is different between the fossil record of whales and these other animals?

6. Explain why the evidence for whale evolution is considered "exceptional." What exception to the general rule does the fossil record of whales represent?

7. Explain what Dr. Clayton Ray, aquatic evolution expert from the Smithsonian Museum, says about how good the evidence is for the evolution of whales?

   Note to student: Dr. Ray was not aware of the new discoveries made by the author detailed later in this chapter.

8. What did Dr. Kevin Padian, Curator of the Museum of Paleontology at the University of California, Berkeley, say about how good the evidence is for the evolution of whales?

   Note to student: Dr. Padian was not aware of the new discoveries made by the author detailed later in this chapter.

9. Explain why, according to Dr. Annalisa Berta (a Professor at San Diego State University who specializes in aquatic mammal evolution), scientists are so excited about the fossil evidence for the evolution of whales.

   Note to student: Dr. Berta was not aware of the new discoveries made by the author detailed later in this chapter.

10. Give the names of the five whale evolutionary intermediates, in order, as proposed by whale evolution experts from the University of Michigan, Ann Arbor.

    Note to student: This was the evidence for whale evolution that Dr. Ray, Dr. Padian, and Dr. Berta were referring to.

    It is very important to memorize these animals by name and in the correct order.

    a.

    b.

    c.

    d.

    e.

11. Scientists who oppose evolution believe the evidence for whale evolution is ____________________.

12. What three land animals do modern evolution scientists say may have evolved into a whale? (page 133).

    a.

    b.

    c.

13. What animal did Darwin think Cetaceans could have evolved from? (See page 40 of Chapter 4.)

14. Why did Charles Darwin remove his idea about bears becoming whales from his book *The Origin of Species?* (See page 40 of Chapter 4.)

15. What do scientists who *oppose* evolution say about the fact that evolution scientists cannot reach a consensus as to which land animal evolved into a whale?

16. Discuss why some evolution scientists chose the cat-like animal, called *Sinonyx*, or the hyena-like animal, called *Pachyaena*, as the land ancestor of whales. What part of the body of these two animals is similar to what part of the body of some extinct whales?

17. What are scientists who *oppose* evolution saying about the tooth evidence of the cat-like animal called *Sinonyx*, the hyena-like animal called *Pachyaena*, and the DNA evidence of hippos?

    a.

    b.

18. Are whales meat eaters or plant eaters?

19. Why did some evolution scientists choose a hippo-like animal as the land ancestor of whales, instead of a cat-like or hyena-like animal? What part of hippos was similar to whales?

1. How is the situation with the whale/hippo/cat/hyena-like animals similar to the hyrax/elephant/sea cow/rhino/horse situation? (See page 61 of Chapter 5.)

2. How do scientists who oppose evolution interpret the similarities in the whale/hippo/cat/hyena-like animals? (See page 60 of Chapter 5.)

3. How do the teeth of whales contradict the evidence of whales evolving from a hippo?

   Note to student: This is the first of two evidences which contradict the idea of hippos being the land mammal that evolved into whales.

4. Why does Dr. Domning, an evolution scientist who specializes in aquatic mammal evolution, think it is "absurd" and "nonsense" to consider hippos as the ancestor of whales? (page 136).

   Note to student: This is the second evidence which contradicts the idea of hippos being the land mammal that evolved into whales.

5. Why was *Ambulocetus* defined as a walking whale?

6. Explain why scientists who oppose evolution believe calling *Ambulocetus* a walking whale is specious (misleading).

7. Describe what body feature of *Ambulocetus* may eliminate it from being on the theoretical evolutionary line to whales and why this would preclude it from being a "walking whale."

8. What is a "fluke"?

9. Which features of *Rodhocetus* suggested to evolution scientists that it was a perfect evolutionary intermediate between a land animal and a whale?

10. How can a scientist tell if an animal had a whale's tail (fluke) or not?

11. What did Dr. Gingerich, a whale evolution expert from the University of Michigan, base his portrayal on of *Rodhocetus* having a fluked tail?

12. Museum drawings and diagrams show *Rodhocetus* with a fluke and front and back flippers. What three parts of the fossil were missing, which should have prevented Dr. Gingerich from interpreting *Rodhocetus* this way?
    a.

    b.

    c.

13. Although some scientists and museum diagrams suggest that ____________________ was on the evolutionary line to modern whales, other evolution scientists disagree with this interpretation.

14. How many of the original five animals on the whale evolution diagram were eliminated during the two interviews with Dr. Gingerich and Dr. Barnes?

15. What do scientists who oppose the theory of evolution have to say about evolution theory, in general, if whale evolution is considered to be one of the best fossil evidences for evolution?

Author's Note: Since the first edition of this book was released, museums have changed their whale evolution displays to correct the serious problems revealed in this chapter. The reader is highly encouraged to now turn to Appendix F: Whale Evolution Update to learn about the latest changes and updates on walking whales.

Purpose of Chapter: To study the fossil record of birds in relation to the theory of evolution. Most evolution scientists suggest that *Archaeopteryx* (and the evolution of birds) is one of the three best fossil proofs for the theory of evolution. This chapter addresses three simple questions: Was *Archaeopteryx* a bird? Or was *Archaeopteryx* an evolutionary intermediate between dinosaurs and birds? Have the evolutionary stages of birds been found as Charles Darwin predicted?

Answer the Discussion Questions. Compare your answers to the next page:

1. What animal do evolution scientists believe birds evolved from? Do you believe this or do you find this idea unbelievable?

2. If birds evolved from dinosaurs, and if the fossil record was nearly complete, what animals should we see in the fossil record demonstrating the evolution of birds from dinosaurs?

3. If birds did not evolve from dinosaurs, and if the fossil record were nearly complete, what animals should we see in the fossil record?

4. How many fossil dinosaurs have been collected by museums and how many fossil birds have been collected by museums?

5. With so many fossil dinosaurs collected and so many fossil birds collected, should we see the evolution of birds from dinosaurs, if, in fact, evolution is true?

Compare your answers:

1. What animal do evolution scientists believe birds evolved from?

   Answer: Many evolution scientists think that birds evolved from dinosaurs. They suggest that a bluebird or hummingbird evolved from a dinosaur.

2. If birds evolved from dinosaurs, and if the fossil record was nearly complete, what animals should we see in the fossil record demonstrating the evolution of birds from dinosaurs?

   Answer: The theory of evolution would predict a scaly reptilian dinosaur slowly developing feathers, slowly developing relatively long arms for wings, and slowly changing the strongest muscles of the body from the back legs for running to the front arms for flapping the wings. If evolution is true, and as the fossil record approaches completeness, many of these animals would be found (see page 75 of Chapter 6).

3. If birds did not evolve from dinosaurs, and if the fossil record were nearly complete, what animals should we see in the fossil record?

   Answer: If birds did not evolve from another animal, such as a dinosaur, then you would see fossil birds and fossil dinosaurs. There would be no intermediate ancestors between the two animal groups.

4. How many fossil dinosaurs have been collected by museums and how many fossil birds have been collected by museums?

   Answer: 100,000 dinosaurs have been collected and 200,000 birds.

5. With this many fossil dinosaurs collected (100,000) and this many fossil birds collected (200,000), should we see the evolution of birds from dinosaurs, if, in fact, evolution is true?

   Answer: Answers may vary but should be thoughtful and logical.

1. In what century was the first fossil specimen of *Archaeopteryx* found?

2. What did evolution scientists believe *Archaeopteryx* proved? Why was it so important to them?

3. List the three best fossil evidences for evolution, according to many evolution scientists.

   a.

   b.

   c.

4. Why do evolution scientists believe *Archaeopteryx* is a missing link between birds and dinosaurs?

5. Where (what country) have all of the *Archaeopteryx* fossils been found?

6. How many *Archaeopteryx* specimens have been found?

7. What are the two possible interpretations of what *Archaeopteryx* looked like?

   a.

   b.

8. Why did scientists make earlier models of *Archaeopteryx* with a scaly reptilian head?

9. Why does Dr. Wellnhofer, an expert and proponent of bird evolution, now believe *Archaeopteryx* had a feathered head similar to modern birds?

10. Why do the newer reconstructions of *Archaeopteryx* look less like a missing link between birds and dinosaurs?

11. What did the head of the older *Archaeopteryx* reconstructions look like?

12. What do newer models of *Archaeopteryx* look like, such as the one at the Chicago Field Museum?

13. What is the significance of finding claws on the wings of *Archaeopteryx*, according to scientists who support evolution?

14. Why do scientists who oppose evolution contend that claws on the wings of *Archaeopteryx* are not proof that birds evolved from dinosaurs?

15. List three other vertebrates (animals with backbones), besides *Archaeopteryx*, that have claws on their wings (page 154).

    a.

    b.

    c.

16. List three types of modern birds that have claws on their wings.

    a.

    b.

    c.

1. Describe the similarities/differences between the tail of dinosaurs, the tail of *Archaeopteryx*, and the tail of modern birds regarding the presence/absence of scales and feathers.

2. Why do evolution scientists believe the tail of *Archaeopteryx* is dinosaur-like while scientists who oppose evolution think the tail of *Archaeopteryx* is bird-like?

3. Why do scientists who support evolution believe the teeth of *Archaeopteryx* prove that birds evolved from dinosaurs?

4. Why do scientists who oppose evolution believe the teeth of *Archaeopteryx* suggest this animal is not linked to dinosaurs? Specifically, how do the teeth of *Archaeopteryx* differ from the teeth of meat-eating dinosaurs?

5. List three birds and how they are similar to other animals, yet according to evolution scientists, are not related to these animals.
   a.
   b.
   c.

6. What would scientists who oppose evolution say proves that birds evolved from dinosaurs?

7. Does *Archaeopteryx* look like a modern bird?

8. Discuss if, superficially, the skeleton of *Archaeopteryx* looks like a modern bird skeleton or not.

9. Can evolution scientists identify which dinosaur theoretically evolved into birds?

10. Which dinosaur is often named the closest relative of *Archaeopteryx*? Why could this dinosaur not be the ancestor of birds (*Archaeopteryx*)?

11. What are the relative ages of *Deinonychus* and *Archaeopteryx*? Which was younger and which was older? Why is this a problem?

    a.

    b.

    c.

12. Scientists who support evolution cannot reach a consensus as to which group of reptiles may have evolved into birds. Name these two different reptile groups.

    a.

    b.

13. Discuss why scientists who oppose evolution suggest that the evolution of birds, and evolution, in general, is "dubious at best."

14. How many fossil birds and how many dinosaurs have been found and why does this undermine the theory of evolution, according to opponents of the theory?

15. List five modern-appearing animals that were found with *Archaeopteryx*. (page 164)

    a.

    b.

    c.

    d.

    e.

16. How do scientists who *oppose* evolution interpret the finding of modern animals alongside *Archaeopteryx*?

17. How do scientists who *support* evolution interpret the finding of modern animals alongside *Archaeopteryx*?

Purpose of Chapter: To study the fossil record of birds in relation to the theory of evolution. Most evolution scientists suggest that the fossil "feathered dinosaurs" found in China in the mid-1990s offer further evidence for the theory of evolution. This chapter addresses these questions: What did the Chinese animals look like? Were the Chinese animals birds or were they half-bird/half-dinosaur? Are the Chinese specimens reliable/authentic or do they need further investigation?

Answer the Discussion Questions. Compare your answers to the next page:

1. Have you ever heard of the "feathered dinosaurs" from China? If so, what did they look like?

2. If you were a scientist (or layperson) trying to decide if evolution is true or not, what *critical* questions would you ask if someone announced they had found a feathered dinosaur that proved, beyond a shadow of a doubt, that birds evolved from dinosaurs?

Compare your answers:

1. Have you ever heard of the "feathered dinosaurs" from China? If so, what did they look like?

   Answer: These fossils are purported by evolution scientists to be dinosaurs with feathers on them, a sort of flying dinosaur, proving evolution.

2. If you were a scientist (or layperson) trying to decide if evolution is true or not, what *critical* questions would you ask if someone announced they had found a feathered dinosaur that proved, beyond a shadow of a doubt, that birds evolved from dinosaurs?

   Answer: You might ask: how much of the fossil did they find? Could it be fake? What level of expertise do the scientists have that are making this claim? Is there consensus among scientists that they are, in fact, feathered dinosaurs? (Or other applicable questions.)

1. When were the Chinese specimens found?

2. In what province of what country were the "feathered dinosaur" specimens found?

3. What do some evolution scientists consider the Chinese specimens to be?

4. What two features of the Chinese specimens lead some evolution scientists to believe they are flightless birds?

5. What two features of the Chinese specimens lead some evolution scientists to believe they are feathered dinosaurs?

6. How large are the Chinese specimens?

7. What does a feather from a modern bird *that can fly* look like, such as a feather from a cardinal or a blue jay?

8. What does a feather from a modern bird *that cannot fly* look like, such as a feather from an ostrich, an emu, or a penguin? List three features of feathers from modern flightless birds that distinguish them from the feathers of modern birds that can fly.

9. Do the feathers from the Chinese specimens appear similar to the feathers of modern flightless birds or do they appear similar to the feathers of modern flying birds? List the three features of the feathers from Chinese specimens to support this answer.

10. Why could the "feathered dinosaurs" not be the ancestors of flying birds, such as *Archaeopteryx*, based on the age of the fossils?

11. What do scientists who oppose evolution say to the idea that the "feathered dinosaurs" from China are the missing link between birds and dinosaurs?

12. Why have some scientists who support evolution raised questions about the authenticity of the Chinese specimens?

13. What does the fossil *Confuciusornis* looks like on a museum shelf and why is this fossil's appearance misleading?

14. What technique did the fossil preparators use to hide the mortar?

15. What bones of the fossil specimen *Confuciusornis* were purposely substituted? What did Dr. Rowe discover concerning the position of one particular bone when he examined the specimen with a CT scan?

   a.

   b.

1. What scientist (and at what university) discovered that *Archaeoraptor liaoningensis* was a purposely faked fossil, a feathered fraud?

2. How did this scientist discover that *Archaeoraptor liaoningensis* was a fraud?

3. *When* (month and year) was the CT scan of *Archaeoraptor liaoningensis* performed (showing it was a problematic fossil) and when was it printed in *National Geographic* magazine (month and year) as being a "feathered dinosaur proving the evolution of birds?" Why was this wrong?

4. How many bones from other animals were fraudulently substituted in *Archaeoraptor liaoningensis*?

5. How did *National Geographic* magazine report on *Archaeoraptor liaoningensis* in the November 1999 issue? What claims did this magazine make about this animal despite evidence that the fossil contained irregularities? Write out three quotes from this issue.

6. How did *National Geographic* magazine do a "disservice" to science, according to Dr. Rowe?

7. How long of an article was the original story about *Archaeoraptor liaoningensis*? How long was the magazine's retraction of the story later? What part of the magazine was the retraction printed in?

   a.

   b.

   c.

8. According to Dr. Rowe, what profession did the perpetrator of the faked fossil *Archaeoraptor liaoningensis* belong to?

9. Describe what *Velociraptor* dinosaurs looked like in the movie Jurassic Park. What do they look like in some museums today? Is this change justified by the fossil evidence?

   a.

   b.

   c.

10. Have scientists ever found fossil evidence of *Velocirpator* dinosaurs having feathers?

11. Write out a list of six major historical examples of *scientists* altering evidence to support the theory of evolution. List them in chronological order from the 19th century to modern times.

    Note to student: Do not include *Confuciusornis* in this list. Also note that the two feathered dinosaur examples are on pages 179 and 174.

    a.

    b.

    c.

    d.

    e.

    f.

12. How have museum artists used fossils to support widely different scientific positions about the evolution of birds from dinosaurs?

13. Which evolution scientist from the Smithsonian Institution publicly chided the National Geographic Society for "making news" and "melodramatic assertions," "editorial propagandizing," "hype," "science fiction," and being "proselytizers of the faith?"

14. Give the species names of the two dinosaurs that were on display at a *National Geographic* exhibit with feathers attached to them. What did Dr. Olson, Curator of Birds at the National Museum of Natural History, have to say about this display (pages 182–183)?

    a.

    b.

    c.

Author's Note: The BBC has uncovered a fossil-faking industry in China in the same area of the world where most of the "feathered dinosaurs" have been found. Equally important, fossils of many modern birds (parrots, ducks, albatross, etc.) have been discovered alongside dinosaurs, seemingly contradicting the theory of evolution. This was reported in the second volume of this book and video series, *Living Fossils*. Turn now to Appendix G: Bird Evolution Update for this most important information!

Purpose of Chapter: To study the fossil record of flowering plants in relation to the theory of evolution. Specifically, this chapter addresses this question: Have the evolutionary stages of flowering plants been found as Charles Darwin predicted? (See pages 75–76 of Chapter 6.)

Answer the Discussion Questions. Compare your answers to the next page:

1. What kinds of trees have flowers in the springtime and are therefore considered flowering trees?

2. Name some plants other than trees that have flowers and are therefore considered flowering plants.

3. What kinds of plants never have flowers and are therefore considered non-flowering plants or trees?

Compare your answers:

1. What kinds of trees have flowers in the springtime and are therefore considered flowering trees?

   Answer: Fruit trees, tulip trees, dogwood trees, etc.

2. Name some plants other than trees that have flowers and are therefore considered flowering plants.

   Answer: Tomatoes, roses, sunflowers, etc.

3. What kinds of plants never have flowers and are therefore considered non-flowering plants or trees?

   Answer: Pine trees or conifers, mosses, cycads.

1. How many plant *species* are alive today?

2. How many *flowering* plant species are alive today?

3. Write out a list of eight examples of flowering plants, including trees.
   a.

   b.

   c.

   d.

   e.

   f.

   g.

   h.

4. What is the scientific term for flowering plants?

5. What was Charles Darwin referring to when he spoke of the "abominable mystery" of flowering plants?

6. What reason did Charles Darwin give for the lack of evidence for plant evolution in the late 1880s?

7. What were plant evolution experts saying in the mid 20th century about plant evolution (70 years after Darwin referred to the lack of evidence for the evolution of flowering plants)?

8. How many fossil plants have been *collected* so far?

9. What is the current evidence for flowering plant evolution, according to *The Encyclopedia of Evolution*? Which of the two models does the fossil record of flowering plants match: the theory of evolution (page 75 of Chapter 6) or the no-evolution model (page 76 of Chapter 6) (page 186)?

   a.

   b.

10. List ten types of plant structures that have been found as fossils, starting with the smallest structure found, such as fossil pollen, and ending with the largest structure found, such as a fossilized whole tree. (Separate each plant part as its own structure.)

    a.

    b.

    c.

    d.

    e.

    f.

    g.

    h.

    i.

    j.

11. Why, according to scientists who *support* evolution, have the intermediate plants demonstrating flowering plant evolution not been found?

12. Why, according to scientists who *oppose* evolution, have the intermediate plants demonstrating flowering plant evolution not been found?

Purpose of Chapter: To discuss the formation of DNA, one of three components necessary for life. According to the theory of evolution, DNA formed naturally, or spontaneously, from chemicals, to create the very first form of life, a single-cell bacterium-like organism. The question is whether this is possible.

Answer the Discussion Questions. Compare your answers to the next page:

1. Have you ever gotten the impression from watching educational television programs that scientists have been able to create life, such as a single-cell organism, in a laboratory using chemicals in a beaker?

2. Conceptually, what does DNA look like?

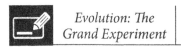

Compare your answers:

1. Have you ever gotten the impression from watching educational television programs that scientists have been able to create life, such as a single-cell organism, in a laboratory using chemicals in a beaker?

   Answer: Students should discuss this. Life (in the form of a single-cell bacterium-like organism) has never been created in a laboratory from chemicals.

2. Conceptually, what does DNA look like?

   Answer: Conceptually, DNA looks like a double spiral helix (twisted ladder). The letters of the DNA (A, C, G, or T) make up the rungs of the twisted ladder.

1. When, according to scientists who support the theory of evolution, did the very first form of life begin?

2. What was the very first form of life, according to scientists who support the theory of evolution?

3. Give the definition of "the origin of life" according to the theory of evolution. (See Glossary.)

4. From what, according to scientists who support the theory of evolution, did all plants and animals living today come?

5. All matter that existed prior to the theoretical spontaneous formation of the first single-cell organism was comprised of ____________________ ____________________, according to the theory of evolution.

6. Have organic molecules been formed successfully in a laboratory?

7. Has a living single-cell organism ever been created successfully in a laboratory using chemicals?

8. Define organic molecules.

9. List four types of organisms living today.

a.

b.

c.

d.

10. List the three necessary components for life to exist in any form.

a.

b.

c.

11. What are the two functions of proteins?

12. What is the function of a cell membrane?

13. What are the two functions of DNA in a living organism?

1. Describe what DNA looks like conceptually.

2. What are the four letters that make up DNA?

3. How many consecutive letters of DNA are needed to instruct the cell to place one amino acid into a protein?

4. What is the formula for calculating the number of DNA letters needed to form one protein 300 amino acids long?

5. How many proteins would be needed for life to begin?

6. What is the formula for calculating the number of DNA letters needed to form twenty proteins, 300 amino acids long, to start life?

7. What is the problem associated with forming DNA strands in a laboratory?

8. What are the odds of blindly choosing *one correct* letter of DNA for assembly into the DNA?

9. What are the odds of blindly choosing *two correct* consecutive letters of DNA in a row? Three letters? Four letters?

   a.

   b.

   c.

10. Which is more likely to occur: accidentally placing 18,000 letters of DNA in proper order OR winning the National Powerball Lottery 365 times in a row?

11. Describe the shape of DNA formed in a laboratory and why this is a problem.

12. Explain why spiraled DNA is so important.

13. How do scientists who *oppose* the theory of evolution respond to experiments showing that functional DNA does not form naturally?

14. How do scientists who *support* the theory of evolution respond to experiments showing that functional DNA does not form naturally?

Purpose of Chapter: To discuss proteins, one of three components necessary for life to begin. According to the theory of evolution, proteins formed naturally, or spontaneously, to create the very first form of life, a single cell bacterium-like organism. The question is whether this is possible.

Optional Project: A chain conveys what proteins conceptually look like. Use a piece of chain maybe one to two feet long, available at most hardware stores, or draw one. The individual links on the protein chain are called amino acids. If the links are big enough, you can write the words amino acid on each link, similar to what is in the book on page 200.

Answer the Discussion Questions. Compare your answers to the next page:

1. What are some good food sources of protein?

2. What do proteins look like?

3. What do proteins in a cell do? What is their function or purpose in a cell?

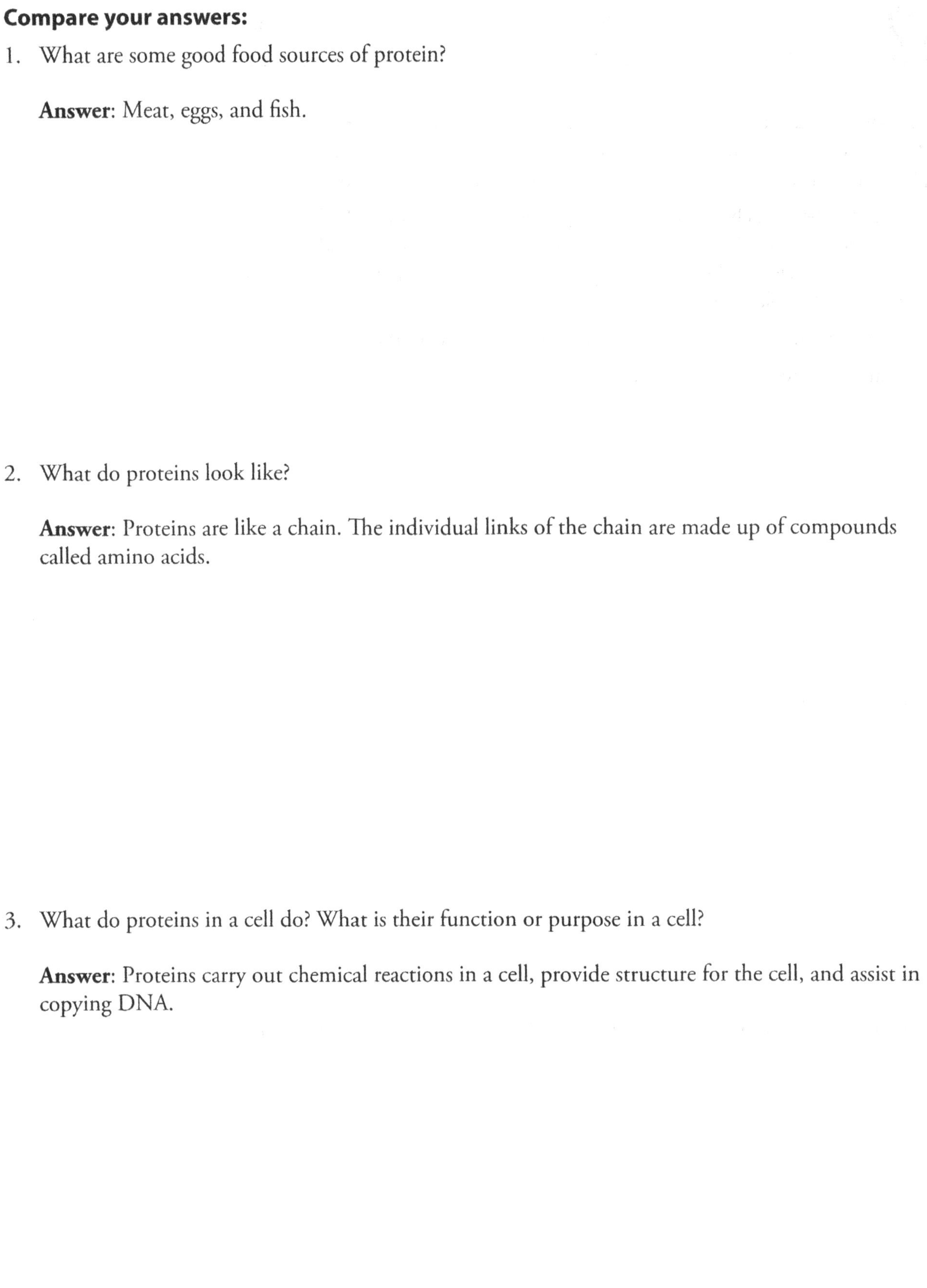

Compare your answers:

1. What are some good food sources of protein?

   Answer: Meat, eggs, and fish.

2. What do proteins look like?

   Answer: Proteins are like a chain. The individual links of the chain are made up of compounds called amino acids.

3. What do proteins in a cell do? What is their function or purpose in a cell?

   Answer: Proteins carry out chemical reactions in a cell, provide structure for the cell, and assist in copying DNA.

1. What are the three basic functions of proteins?
   a.

   b.

   c.

2. Do proteins form naturally?

3. What makes up the individual links of a protein chain?

4. Are proteins necessary for life?

5. Why is it a problem to believe that the first life formed in the ocean, as suggested by some scientists who support the theory of evolution?

6. Has anyone has ever witnessed a protein forming naturally, according to biochemist and evolution opponent Dr. Duane Gish?

7. Explain what a *proteinoid* is by writing out the first line of the definition from the Glossary.

8. Explain, conceptually speaking, what both a protein and a proteinoid look like.

9. How did proteinoids become proteins, according to scientists who support evolution?

10. Modern evolution scientists do not believe proteinoids formed in the ocean because water prevents this process. Where did proteinoids form then, according to scientists who support evolution?

11. Some scientists who *support* evolution have questioned the scenario of proteinoids forming on the side of a heated volcano and then later forming the very first single-cell bacterium-like organism in the ocean. These scientists ask an even more basic question. What is the question they ask?

12. Some scientists who *oppose* evolution have questioned the scenario of proteinoids starting life and think the whole idea is preposterous. What three points do these scientists make to support their position?

    a.

    b.

    c.

13. How many different types of *amino acids* are found in living organisms today, such as plants, animals, and bacteria?

14. What are the consequences of placing one wrong amino acid in a protein chain?

15. How many unique proteins are in the simplest bacterium *living today*?

16. How many unique proteins would be necessary to begin life, in the most rudimentary theoretical *first* single-cell organism or bacteria-like organism, according to scientists who support the theory of evolution?

17. How many amino acids have to line up in a protein chain, in the correct order, to begin the theoretical first form of life, i.e., a single cell bacterium-like organism?

18. What is the *formula* for calculating the number of amino acids needed to assemble into a protein for life to begin?

19. What two reasons does Dr. Strickberger give for questioning the possibility of a single protein forming in the correct (amino acid) sequence? (Dr. Strickberger is a scientist who supports evolution and author of a popular college evolution textbook.)

    a.

    b.

20. What do scientists who oppose evolution say about the fact that proteins (and proteinoids) do not form naturally?

Purpose of Chapter: To discuss amino acids, a component of proteins. According to the theory of evolution, amino acids would have had to form spontaneously to produce the first proteins (or proteinoids) necessary for the theoretical first form of life, a single-cell bacterium-like organism. This chapter looks at the problems of forming amino acids naturally (or spontaneously).

Note: Again, it is suggested that a chain be used (introduced in Chapter 18) as a model for the discussion below. Conceptually, the chain represents what a protein looks like. Amino acids make up the individual links of protein chains.

Answer the Discussion Questions. Compare your answers to the next page:

1. Foods rich in protein would be eggs, fish, and meat. Conceptually, what do proteins look like?

2. If proteins are needed for life to exist, why would amino acids also be necessary for life to begin or exist?

3. Do amino acids form spontaneously (or without the help of a chemist)?

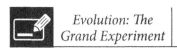

Compare your answers:

1. Foods rich in protein would be eggs, fish, and meat. Conceptually, what do proteins look like?

   Answer: Use the chain or draw a chain with links on the board. The chain represents what a protein looks like. The individual links of the protein chain are amino acids.

2. If proteins are needed for life to exist, why would amino acids also be necessary for life to begin or exist?

   Answer: Amino acids make up a protein chain, so both are needed for life to exist.

3. Do amino acids form spontaneously (or without the help of a chemist)?

   Answer: Answers may vary but the student should give reasons for their answer. (The answer is in Chapter 19.)

1. When, according to the theory of evolution, did life begin (in the form of the very first single-cell bacterium-like organism)?

2. Proteins are necessary for life. Therefore, __________ __________, which make up proteins, would have had to form naturally if evolution is true.

3. Who (first) produced amino acids in the laboratory and in what year?

   a.

   b.

4. List the components (or parts) of the apparatus Stanley Miller used to form amino acids in his laboratory under "natural conditions."

   a. ______________________ filled with water and chemicals.

   b. ______________________ ____________________ to provide an electrical charge.

   c. ______________________ device using a cold-water condenser for separating amino acids from the tungsten electrode after they formed.

   d. ______________________ container for collecting amino acids after distillation/condensation.

5. What did Dr. Miller theorize his experiment represented?

6. What natural occurrence have Dr. Miller and other evolution scientists suggested the tungsten electrode used in his experiment (to create a spark) represented?

1. What criticisms of Miller's experiment have been given by scientists who oppose evolution?
   a.
   b.
   c.
   d.
   e.
2. What challenge has been given by scientists who oppose evolution regarding the use of sophisticated equipment to produce amino acids?
3. Stanley Miller removed all traces of ____________________ from his apparatus before beginning his experiment.
4. Explain what would have happened if trace amounts of oxygen were present in Miller's equipment.
5. What, according to evolution scientists, did the earth have to lack in order to be consistent with Dr. Miller's experiment?
6. What evidence is there to suggest oxygen *was* present on the earth around 4 billion years ago?
7. If oxygen was present on the earth, would that nullify Stanley Miller's experiment?

8. Explain the "catch-22" (or paradox) concerning the presence or absence of destructive oxygen vs. protective ozone at the time of the formation of the theoretical first single-cell organism.

9. What kind of amino acids did Stanley Miller's apparatus produce: left or right-handed?

10. What problems are there with right-handed amino acids and why is this a problem for Stanley Miller's experiment?

11. How many amino acid types are used by living plants, animals, and bacteria today?

12. Did Stanley Miller's apparatus produce all of the 20 needed types of amino acids used in living organisms today?

13. Explain the "catch-22" associated with Stanley Miller's experiment relating to water and the formation of life.

14. Has a cell membrane ever been shown to form naturally?

15. Have scientists ever been able to create a living organism in a laboratory only using a mixture of chemicals?

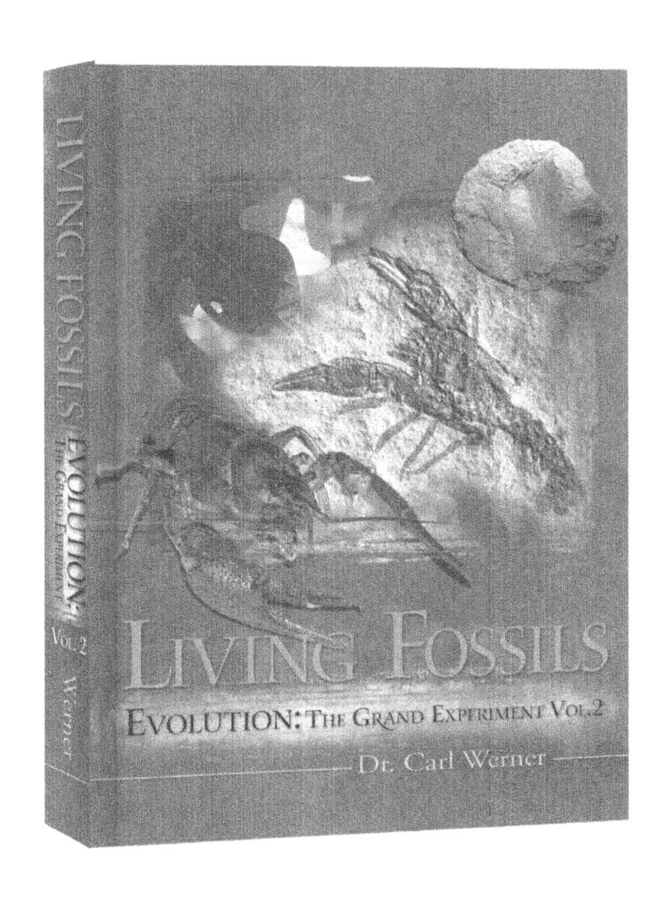

Worksheets

for Use with

Living Fossils

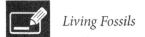

Purpose of Chapter: Learn how the author was challenged by a college classmate to question his belief in evolution. What was the author asked that made him begin to question evolution?

Answer the Discussion Questions. Compare your answers to the next page:

1. If a teacher at college tells you something is true, is it necessarily true? Should you believe everything you are taught?

2. If an idea is taught by scientists for 100 years, does it mean it is necessarily true?

Compare your answers:

1. If a teacher at college tells you something is true, is it necessarily true? Should you believe everything you are taught?

   Answer: Teachers are human and they can be wrong. Anytime you are taught something that does not make sense, you should be hesitant to accept it without discussing it with others and diligently scrutinizing it.

2. If an idea is taught by scientists for 100 years, does it mean it is necessarily true?

   Answer: Scientists have been wrong about scientific ideas in the past. Sometimes acceptance of one of these incorrect ideas can last for *decades, centuries,* and even *millennia.* Here are some examples:

   For thousands of years, scientists taught that the *earth* was the center of our planetary system. This is incorrect. The sun is the center of our planetary system. Scientists taught that mice came from dirty underwear and maggots came from rotting meat. Again, these ideas were completely wrong but were taught for over 2,000 years.

   Just because a group of scientists say something is true, it is not *necessarily* true.

1. Write out an outline of Darwin's theory of evolution using only 10 words: (see pages 2 and 4 in book)
   a. ____________________ (became)
   b. ____________________ (became)
   c. ____________________ (became)
   d. ____________________ (became)
   e. ____________________ (became)
   f. ____________________ (became)
   g. ____________________ (became)
   h. ____________________ (became)
   i. ____________________ (became)
   j. ____________________

2. What is one evidence for evolution the author heard from his college professor in 1977 that made him "rapidly accept evolution?"

3. What is "Ontogeny Recapitulates Phylogeny?"

4. According to the 1990 *Encyclopedia of Evolution* (a pro-evolution book), what wrong had Haeckel engaged in in his famous embryo illustrations used to demonstrate Ontogeny Recapitulates Phylogeny?

5. List three errors of Dr. Haeckel's "Ontogeny Recapitulates Phylogeny."
   a.
   b.
   c.

6. What were the years Dr. Ernst Haeckel lived and died? Given that Charles Darwin published his theory *The Origin of Species* in 1859, was Dr. Haeckel alive when Darwin wrote *The Origin of Species*?

   a.

   b.

7. In what year was the author taught "Ontogeny Recapitulates Phylogeny" in college? When was "Ontogeny Recapitulates Phylogeny" shown to be false?

   a.

   b.

8. What were the four questions posed to the author in college which made him begin to question evolution?

   a.

   b.

   c.

   d.

9. What prize did Dr. David Gross receive? What year did he receive it?

   a.

   b.

10. Tell what two words Dr. David Gross used while describing the laws of physics in the big bang; "We find ___________ _____________."

Purpose of Chapter: Discuss how the theory of evolution or any other scientific theory could be wrong. Sometimes scientists make mistakes.

Answer the Discussion Questions. Compare your answers to the next page:

1. Is it possible for a scientist to make a mistake? Give an example.

2. Have scientists made any mistakes in your lifetime?

3. If a majority of scientists believe in something, should you accept what they believe?

Compare your answers:

1. Is it possible for a scientist to make a mistake? Give an example.

   Answer: Yes. Scientists used to (incorrectly) believe the earth was the center of our solar system. (Galileo and others later showed that the sun was the center of our solar system.) Also, in the past, scientists thought that the earth was flat. We now know, of course, that the earth is round, like a ball.

2. Have scientists made any mistakes in your lifetime?

   Answer: Yes. Recent examples include cold fusion, the belief that certain foods cause certain diseases, etc.

3. If a majority of scientists believe in something, should you accept what they believe?

   Answer: No, not necessarily. Scientists can be wrong.

1. Give two historical examples of scientists believing in a scientific idea which later was shown to be wrong.

   a.

   b.

2. What are the three evidences against evolution which caused the author to question the theory of evolution?

   a.

   b.

   c.

3. What is the scientific method?

4. How does a scientist test a theory or idea?

5. What are the two basic components of the theory of evolution?

   a.

   b.

6. Using the scientific method, the author tried to test the theory of evolution by taking the opposite stance. He predicted two things should be true if evolution was not true. What were these two predictions?

   a.

   b.

7. Scientists who support evolution assign three names to rock layers that contain dinosaurs. List these names.

   a.

   b.

   c.

8. What is a "living fossil?"

9. What are the four most well-recognized examples of living fossils from the dinosaur layers?

   a.

   b.

   c.

   d.

10. How many miles did the author travel to *test* evolution compared to how far Charles Darwin traveled to *prove* evolution?

11. What question did the author plan to ask to test evolution?

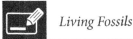

Purpose of Chapter: To teach about the subjective nature of assigning species names to fossils and to describe how this process could falsely lead one to conclude that animals and plants changed dramatically over time. This chapter introduces a few key concepts that will be used throughout the semester. Understand the definition of "species" and what the author predicted he would find if evolution was not true.

Answer the Discussion Questions. Compare your answers to the next page:

1. What is the definition of a species? How can you tell if two animals are the same species or not? Give some examples of same species and different species.

2. A horse can reproduce with a donkey and produce a hybrid animal, a mule, but the mule is unable to reproduce. It is sterile. Knowing these facts, are a horse and a donkey the same species?

3. What is the species name for a dog? For example, what is the species names for a beagle or a German Shepherd?

4. Do you think it is possible for a scientist to be wrong when he or she assigns a species name to a fossil? Why?

Compare your answers:

1. What is the definition of a species? How could you tell if two animals are the same species or not? Give some examples of same species and different species.

   Answer: You can tell if two animals are the same species if they can reproduce and the offspring of the animals are also able to reproduce. Two dogs are the same species because they can reproduce and have puppies and when the puppies get older they can also reproduce. A cat and a dog cannot reproduce, so they are not the same species.

2. A horse can reproduce with a donkey and produce a hybrid animal, a mule, but the mule is unable to reproduce. It is sterile. Knowing these facts, are a horse and a donkey the same species?

   Answer: You can tell if two animals are the same species if they can reproduce and the offspring are also able to reproduce. Since a the offspring of a horse and a donkey, a mule, cannot reproduce, then by definition, a horse and a donkey are different species. (A horse has 64 chromosomes and a donkey only has 62 chromosomes.)

3. What is the species name for a dog? For example, what is the species names for a beagle or a German Shepherd?

   Answer: *Canis familiaris* is the species name for all breeds of domestic dogs.

4. Do you think it is possible for a scientist to be wrong when he or she assigns a species name to a fossil? Why?

   Answer: Scientists cannot test fossils for the ability to produce fertile offspring. Because of this, there is no way to tell if the species name they assign is right or wrong.

1. Write out the kingdom, phylum, class, order, family, genus, and species names for a dog, a horse, and a human according to evolution scientists.

   Dog

   a. Kingdom:

   b. Phylum:

   c. Class:

   d. Order:

   e. Family:

   f. Genus:

   g. Species:

   Horse

   h. Kingdom:

   i. Phylum:

   j. Class:

   k. Order:

   l. Family:

   m. Genus:

   n. Species:

   Human

   o. Kingdom:

   p. Phylum:

   q. Class:

   r. Order:

   s. Family:

   t. Genus:

   u. Species:

2. Arrange these groups in order from biggest group to smallest: genus, order, kingdom, phylum, family, class, and species.

3. What word is used for plants instead of the word phylum?

4. A scientific name is formed for an animal based on its ________________ and ________________ names, and a species name is written in ________________.

5. Give the common names for:

   a. *Homo sapiens* -

   b. *Equus caballus* -

   c. *Canis familiaris* -

   d. *Canis lupus* -

6. What is the simplest, most straightforward definition of a species?

7. Horses and ________________ are not the same species.

8. Give the number of chromosomes each animal has (page 18).

   a. Horse-

   b. Donkey-

   c. Mule-

9. Explain why the horse and the donkey are not considered the same species even though they can mate and produce a mule.

10. Explain why a cat and a dog are not the same species.

11. There are _________________ variations within any one species, such as the variations in skulls, jaws, and teeth of dogs, shape of skulls and bones in a human being, shape of shells of American Oysters, shape of the leaves of a sassafras tree, and the shape of the leaves on a star magnolia tree.

12. A Beagle, a Doberman, and a German Shepherd are the same ____________________ (*Canis familiaris*), even though they look very different.

13. List the four potential reasons that a scientist may assign an incorrect species name to a fossil.

    a.

    b.

    c.

    d.

14. Evolution scientists mistakenly assigned human beings to more than ____________________ different species.

15. Biologically, there is only one ____________________ of humans.

16. A fossil magnolia leaf found with a dinosaur looks _______________________ to a modern magnolia leaf, yet it was assigned a different species name.

17. Write out what the author predicted he would find in dinosaur rock layers if evolution was not true.

Purpose of Chapter: To test the author's idea that if evolution is *not* true, modern-appearing echinoderms should be found with dinosaurs.

Answer the Discussion Questions. Compare your answers to the next page:

1. Is it acceptable to question a scientist's conclusions?

2. If a scientist assigns a fossil starfish a completely different genus or species name than a modern starfish, yet they look nearly the same, is it correct to question this change of genus and species names?

Compare your answers:

1. Is it acceptable to question a scientist's conclusions?

   Answer: Scientists have been wrong about many things in the past. Because of this, you should feel free to openly question anything a scientist proposes, especially if you cannot verify it for yourself or if it does not make sense.

2. If a scientist assigns a fossil starfish a completely different genus or species name than a modern starfish, yet they look nearly the same, is it correct to question this change of genus and species names?

   Answer: Scientists have a long history of assigning living and fossil plants and animals to the wrong species. The species assigned to any animal or plant is not *necessarily* accurate.

1. List five living animals belonging to the Echinoderm phylum.
   a.
   b.
   c.
   d.
   e.
2. Humans have two-sided symmetry, a right and a left side. What kind of symmetry do echinoderms have?
3. Describe the two types of sea urchins.
   a.
   b.
4. What are the two types of crinoids?
   a.
   b.
5. List the five major classes of echinoderms living today.
   a.
   b.
   c.
   d.
   e.

6. How many of the five major classes of echinoderms living today did the author find in dinosaur rock layers?

7. The author predicted that if evolution was not true, he would find modern types of echinoderms in dinosaur rock layers. Was his prediction validated?

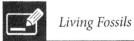

Purpose of Chapter: To test the author's idea that if evolution is *not* true, modern-appearing aquatic arthropods should be found with dinosaurs.

Answer the Discussion Questions. Compare your answers to the next page:

1. What did the author of this book predict he would find in dinosaur rock layers if evolution was not true?

2. Can you be absolutely sure that any species name assigned to a fossil is correct just because a scientist assigned that name?

3. Insects, spiders, lobsters, crayfish, crabs, and scorpions are all classified as "arthropods." What features do all of these animals have in common? What is the definition of an arthropod?

Compare your answers:

1. What did the author of this book predict he would find in dinosaur rock layers if evolution was not true?

   Answer: Modern species of animals and plants should be found in dinosaur rock layers.

2. Can you be absolutely sure that any species name assigned to a fossil is correct just because a scientist assigned that name?

   Answer: No, scientists can be wrong choosing species names. When dealing with fossils this is especially true since no one can test for reproduction of a species.

3. Insects, spiders, lobsters, crayfish, crabs, and scorpions are all classified as "arthropods." What features do all of these animals have in common? What is the definition of an arthropod?

   Answer: Arthropods have an outer armor called an *exoskeleton, segmented bodies*, and *jointed legs.*

1. List the three features an animal must have to be an arthropod.

   a.

   b.

   c.

2. What is the formal phylum name of the phylum which includes starfish? (See page 29.) What is the formal phylum name which includes crayfish (page 45)?

   a.

   b.

3. Name the bird and the dinosaur found in Solnhofen, Germany, in Jurassic rock layers.

   a.

   b.

4. Write out five types of modern-appearing aquatic arthropods also found in Solnhofen, Germany. (Hint, look for the word "Solnhofen" in the caption of the fossils on pages 48-54.)

   a.

   b.

   c.

   d.

   e.

5. What is the aquatic arthropod group living today not found in Solnhofen, Germany, but has been found in dinosaur fossil layers?

6. All of the _______________ major types of aquatic arthropods living today have been found as fossils in the dinosaur layers, too.

7. Decide if the photographs in this chapter support the author's prediction that if evolution was *not* true he would find modern-appearing aquatic arthropods in dinosaur layers.

Purpose of Chapter: To test the author's idea that if evolution was *not* true, modern-appearing *land* arthropods should be found with dinosaurs.

Answer the Discussion Questions. Compare your answers to the next page:

1. What is an arthropod?

2. Give several examples of land arthropods (Chapter 6) and aquatic arthropods (Chapter 5).

3. What did the author predict he would find in dinosaur rock layers regarding land arthropods if evolution was *not* true?

Compare your answers:

1. What is an arthropod?

   Answer: An arthropod, such as an insect, is an animal with an outer covering (called an exoskeleton), jointed legs, and a segmented body.

2. Give several examples of land arthropods (Chapter 6) and aquatic arthropods (Chapter 5).

   Answer: Land arthropods include insects, scorpions, and spiders. Aquatic arthropods include shrimp, lobster, crab, and crayfish.

3. What did the author predict he would find in dinosaur rock layers regarding land arthropods if evolution was *not* true?

   Answer: He predicted he would find modern-appearing land arthropods such as insects, scorpions, and spiders in dinosaur rock layers.

1. Name five land arthropods (page 57).
   a.

   b.

   c.

   d.

   e.

2. What percentage of the major insect orders living today were also found in dinosaur rock layers?

3. Name the two major types of Myriapods living today. How many of the two major Myriapod classes were found in dinosaur rock layers?
   a.

   b.

   c.

4. What percentage of the major Arachnid orders living today were also found in dinosaur rock layers?

5. What did the author's experiment predict he would find regarding land arthropods if evolution was not true? (See page 28.)

6. Over time, the author began to trust less in the species _______________ assigned to the fossils and relied more on his own judgement by comparing photographs of the fossils with the living organisms. (See page 28.)

7. Decide if the author's prediction regarding land arthropods came true or not.

Purpose of Chapter: To test the author's idea that if evolution was *not* true, modern-appearing bivalve shellfish should be found in dinosaur rock layers.

Answer the Discussion Questions. Compare your answers to the next page:

1. How many shells do clams have? How many shells do snails have?

2. Can you think of any common shellfish that have two shells?

3. This is a review question: In which animal phylum are starfish?

4. This is a review question: Which animal phylum includes shrimp and insects?

5. In which animal phylum do clams and snails belong?

6. What did the author predict he would find concerning shellfish in the dinosaur rock layers if evolution was *not* true?

Compare your answers:

1. How many shells do clams have? How many shells do snails have?

   Answer: Snails have one shell, while clams have two shells.

2. Can you think of any common shellfish that have two shells?

   Answer: Clams, scallops, oysters, mussels, and many others.

3. This is a review question: In which animal phylum are starfish?

   Answer: Echinoderms or Phylum Echinodermata.

4. This is a review question: Which animal phylum includes shrimp and insects?

   Answer: Arthropod or Phylum Arthropoda.

5. In which animal phylum do clams and snails belong?

   Answer: Phylum Mollusca. You will learn about this in the chapter.

6. What did the author predict he would find concerning shellfish in the dinosaur rock layers if evolution was *not* true?

   Answer: He predicted he would find modern-appearing shellfish in the dinosaur layers.

1. Snails have one shell and ____________________ shellfish have two shells.

2. List the four common bivalve shellfish types living today.
   a.
   b.
   c.
   d.

3. What shell is pictured on the famous Shell Oil Company sign?

4. What are the two parts of the scallop shell? Which part breaks off frequently?
   a.
   b.
   c.

5. List the five common bivalve shellfish that have been found in dinosaur rock layers.
   a.
   b.
   c.
   d.
   e.

6. Give the phyla for: (see pages 71, 77, 83)

   a. bivalve shellfish-

   b. snails-

   c. chambered nautilus-

   d. sea cradles-

   e. tusk shells-

7. The author predicted he would find modern-appearing bivalve shellfish in the dinosaur layers if evolution was not true. Did his prediction come true?

Purpose of Chapter: To test the author's idea that if evolution was *not* true, modern-appearing snails should be found in dinosaur rock layers.

Answer the Discussion Questions. Compare your answers to the next page:

1. Let's review the animal phyla groups we have studied so far. First, we looked at starfish and sea urchins in Chapter 4. What phylum are they in?

2. In Chapters 5 and 6 we looked at shrimp, lobsters, insects, and spiders. What phylum are they in?

3. In Chapter 7 we looked at clams, scallops, and other bivalves. What phylum are they in?

4. So far we have looked at three phyla (out of the 7 major phyla) of animals. The author of this book predicted that he would find modern-appearing animals in dinosaur layers if evolution was not true. So far, has he found modern-appearing examples from these first three phyla we looked at: echinoderms, arthropods and mollusks?

Compare your answers:

1. Let's review the animal phyla groups we have studied so far. First, we looked at starfish and sea urchins in Chapter 4. What phylum are they in?

   Answer: Echinoderms or Phylum Echinodermata.

2. In Chapters 5 and 6 we looked at shrimp, lobsters, insects, and spiders. What phylum are they in?

   Answer: Arthropods or Phylum Arthropoda.

3. In Chapter 7 we looked at clams, scallops, and other bivalves. What phylum are they in?

   Answer: Molluscs or Phylum Mollusca.

4. So far we have looked at three phyla (out of the 7 major phyla) of animals. The author of this book predicted that he would find modern-appearing animals in dinosaur layers if evolution was not true. So far, has he found modern-appearing examples from these first three phyla we looked at: echinoderms, arthropods and mollusks?

   Answer: Yes.

1. What phylum do snails belong to and what phylum includes bivalves? (See pages 71 and 77.)
   a.

   b.

2. How many shells does a snail have and how many shells does a bivalve shellfish have?
   a.

   b.

3. At what national monument in the United States were modern appearing *freshwater* snails found in dinosaur rock layers?

4. What two types of dinosaurs are found at Dinosaur National Monument?
   a.

   b.

5. At what national monument in the United States were modern appearing *saltwater* snails found in dinosaur rock layers?

6. What type of dinosaur was found with a fossilized snail at Petrified Forest National Park?

7. Modern-appearing slit shells and moon snails were found in ____________________ dinosaur rock layers.

8. The author predicted he would find modern-appearing snails in dinosaur rock layers if evolution was *not* true. Did he find them?

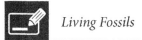

Purpose of Chapter: To test the author's idea that if evolution was *not* true, other types of modern-appearing shellfish should be found with dinosaurs.

Answer the Discussion Questions. Compare your answers to the next page:

1. Have you ever seen a tusk shell? What do they look like?

2. Have you ever seen a nautilus shell? What do they look like?

3. Have you ever seen a lamp shell or sea cradle?

Compare your answers:

1. Have you ever seen a tusk shell? What do they look like?

   Answer: Answers will vary. Tusk shells are about the size of your finger and they look like an elephant tusk.

2. Have you ever seen a nautilus shell? What do they look like?

   Answer: Answers will vary. A nautilus looks like a big snail, but it has chambers.

3. Have you ever seen a lamp shell or sea cradle?

   Answer: Answers will vary. All of these shells can be seen when you read Chapter 9.

1. The ____________________ is not a snail because it has chambers. Snails do not have chambers.

2. The snail and the nautilus are members of the Phylum ____________________.

3. In a nautilus, the tube-like ______________________ cuts through the center of each chamber.

4. Tusk shells were found fossilized in dinosaur rock layers in the United States and look like _________________ tusks, but are only inches long.

5. Tusk shells are members of the Phylum ________________.

6. Sea cradles are members of the Phylum ________________.

7. What are sea cradles and how many interlocking plates do they possess?

8. Sea cradles were found in ____________________ ____________________ layers.

9. The author predicted that if evolution was *not* true, he should find modern-appearing mollusks in _____________________ ____________________ layers. (See page 28.)

10. Fill in the five major classes of shellfish living today in the Phylum Mollusca, which have also been found in dinosaur rock layers:

    a. Bivalves (such as scallops, clams, oysters, mussels) = Class __________________

    b. Snails (such as freshwater snails, saltwater snails, slit shells) = Class __________________

    c. Chambered shellfish (such as the nautilus) = Class __________________

    d. Tusk shells = Class __________________

    e. Sea cradles = Class __________________

11. Fill in this summary table for the Phylum Mollusca.

Phylum	**Class**	**Example**	**Living Today?**	**Found with Dinos?**
Mollusca	Bivalvia	Clam		
Mollusca	Gastropoda	Freshwater Snail		
Mollusca	Cephalopoda	Nautilus		
Mollusca	Scaphoda	Tusk shell		
Mollusca	Polyplacophora	Sea Cradle		

12. Does the table above support the author's prediction that if evolution was *not* true, he would find modern types of shellfish with dinosaurs?

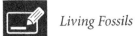

Purpose of Chapter: To test the author's idea that if evolution was *not* true, modern-appearing worms should be found with dinosaurs.

Note: There are approximately 34 animal phyla living today. (Some experts say there are 34 animal phyla, others 40, others somewhere between. Each expert groups animals differently.)

This book covers only the seven common animal phyla. Most people have never seen examples from any of the other 27 animal phyla. The author of this book has focused only on common animal phyla, animals that everyone would recognize. For example, this chapter deals with the most familiar worm, the earthworm (Phylum Annelida). The author has purposely left out of this book the following 13 unfamiliar worm phyla groups:

Other 13 Worm Phyla	Examples
1. Pogonophora	Beard Worms
2. Chaetognatha	Arrow Worms
3. Nemertina	Ribbon Worms
4. Gnathostomulida	Sand Worms
5. Priapulida	Phallus Worms
6. Sipuncula	Peanut Worms
7. Echiura	Spoon Worms
8. Platyhelminthes	Flat Worms
9. Pentastoma	Tongue Worms
10. Nematomorpha	Horsehair Worms
11. Kinorhyncha	Spiny-crown Worms
12. Acanthocephala	Spiny-headed Worms
13. Nematoda	Round Worms (Not earthworms)

Because the author wants to keep the information simple for all to learn, he has focused on what he calls "the seven *major* animal phyla," which are:

1. Phylum Echinodermata (starfish group)
2. Phylum Arthropoda (insects, spiders, shrimp, lobsters, etc.)
3. Phylum Mollusca (shellfish)
4. Phylum Annelida (earthworms)
5. Phylum Porifera (sponges)
6. Phylum Cnidaria (corals)
7. Phylum Chordata (vertebrates such as fish, mammals, amphibians, reptiles)

Answer the Discussion Questions. Compare your answers to the next page:

1. Do you think that earthworms were alive at the same time that *T. rex* was alive?

2. To what phylum do earthworms belong?

Compare your answers:

1. Do you think that earthworms were alive at the same time that *T. rex* was alive?

   Answer: Answers may vary. (The answer is in Chapter 10.)

2. To what phylum do earthworms belong?

   Answer: Annelida. (The answer is in Chapter 10.)

1. Give the official phylum name for segmented worms and give two examples of segmented worms (pages 89, 91-92).

   a.

   b.

   c.

2. There are only two classes of segmented worms living today (Polychaeta and Oligochaeta) and both have been found as fossils in ____________________ rock layers. Tube worms are members of the class Polychaeta. Earthworms are members of the other class, ______________________.

3. Where do tube worms live today?

4. There are 14 worm phyla. This book only covers the most familiar type of worm, the segmented worms, Phylum Annelida, which includes earthworms and tube worms. Tell if you have ever seen examples of the other 13 obscure worm phyla that were not included in this book.

Worm Phyla	Worms	Included in This Book?	Have you seen?
1. Annelida	Earthworms	Yes, Chapter 10	
2. Pogonophora	Beard Worms	No	
3. Chaetognatha	Arrow Worms	No	
4. Nemertina	Ribbon Worms	No	
5. Gnathostomulida	Sand Worms	No	
6. Priapulida	Phallus Worms	No	
7. Sipuncula	Peanut Worms	No	
8. Echiura	Spoon Worms	No	
9. Platyhelminthes	Flat Worms	No	
10. Pentastoma	Tongue Worms	No	
11. Nematomorpha	Horsehair Worms	No	
12. Kinorhyncha	Spiny-crown Worms	No	
13. Acanthocephala	Spiny-headed Worms	No	
14. Nematoda	Roundworms (not earthworms)	No	

Purpose of Chapter: To test the author's idea that if evolution was *not* true, modern-appearing corals and sponges should be found with dinosaurs.

Answer the Discussion Questions. Compare your answers to the next page:

1. Corals and sponges live in oceans today. Are corals and sponges plants? Animals? Fungi?

2. Have you swum in the ocean and seen a living coral or sponge?

3. Animals are divided into two categories, vertebrates and invertebrates. Vertebrates, such as mammals and reptiles, have a backbone with vertebrae. Invertebrates, such as starfish, do not have a backbone. Are sponges and corals vertebrates or invertebrates?

Compare your answers:

1. Corals and sponges live in the oceans today. Are corals and sponges plants? Animals? Fungi?

   Answer: Although corals and sponges usually are stationary like plants, they are animals that live in the ocean.

2. Have you swum in the ocean and seen a living coral or sponge?

   Answer: Answers will vary.

3. Animals are divided into two categories, vertebrates and invertebrates. Vertebrates, such as mammals and reptiles, have a backbone with vertebrae. Invertebrates, such as starfish, do not have a backbone. Are sponges and corals vertebrates or invertebrates?

   Answer: Sponges and corals are *invertebrates*, since they do not have a backbone or spinal cord.

1. Sponges are members of the Phylum ________________ and corals are members of the Phylum ________________ (also called Phyla Coelenterata).

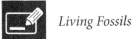

2. The author predicted that if evolution is not true, he would find modern-appearing corals and sponges in ___________________ rock layers.

3. List the three classes of sponges living today that were also living at the time of the dinosaurs: (See pages 94, 95, 98, and Bibliography Chapter 11, Footnotes 1 and 2.)
   a. Bony sponges-Class ________________
   b. Glass sponges-Class ________________
   c. Spongin-Class ________________

4. Both groups of corals living today, __________________ corals and __________________ corals, lived at the time of the dinosaurs. (See pages 96-98, and Bibliography Chapter 11, Footnote 3.)

5. There are approximately __________ animal phyla living today.

6. This book covers only the seven common animal phyla. Most people have never *seen* any of these other obscure animal phyla. The author purposely omitted the 13 obscure worm phyla listed in Chapter 10 and these other 14 unfamiliar non-worm invertebrate phyla groups from his "major phyla list."

Other Phyla	Example	Have you ever seen any of these in nature?
1. Placozoa	Placozoa	
2. Mesozoa	Mesozoa	
3. Gastrotricha	Gastrotrichs	
4. Rotifera	Rotifers	
5. Tardigrada	Water Bears	
6. Onychophora	Peripatus	
7. Loricifera	Brush Heads	
8. Cycliophora	Cycliophorans	
9. Entoprocta	Marine Mats	
10. Ectoprocta	Bryozoans	
11. Phoronida	Phoronans	
12. Ctenophora	Comb Jellies	
13. Hemichordata	Hemichordates	
14. Brachiopoda	Lamp Shells (See pages 83, 87.)	

Purpose of Chapter: To test the author's idea that if evolution was *not* true, modern-appearing bony fish should be found with dinosaurs.

Answer the Discussion Questions. Compare your answers to the next page:

1. If evolution is true and animals changed over time, would you expect to find modern animals, such as modern fish, in dinosaur rock layers?

2. If all the animals were created at one time, such as humans, dinosaurs, catfish, salmon, and sharks, as in the biblical account of creation, would you expect to find *modern* fish in dinosaur rock layers?

3. Did modern fish live at the same time as *T. rex* (as the theory of creation predicts) or did only strange and extinct fish live with *T. Rex* (as the theory of evolution predicts)?

Compare your answers:

1. If evolution is true and animals changed over time, would you expect to find modern animals, such as modern fish, in dinosaur rock layers?

   Answer: If animals have changed over time as evolution predicts, modern animals, including modern fish, should *not* be found with dinosaur fossils.

2. If all the animals were created at one time, such as humans, dinosaurs, catfish, salmon, and sharks, as in the biblical account of creation, would you expect to find *modern* fish in dinosaur rock layers?

   Answer: The answer is yes. If all the animals were created at the same time, you would expect to find modern fish with dinosaurs.

3. Did modern fish live at the same time as *T. rex* (as the theory of creation predicts) or did only strange and extinct fish live with *T. Rex* (as the theory of evolution predicts)?

   Answer: You will find the answer in this lesson.

1. Phylum Chordata includes all animals with a ________________ ______________ and includes fish, amphibians, reptiles, birds, and mammals.

2. Animals with ______________________ (also called a vertebrae) are called vertebrates. All vertebrates have a spinal cord and belong to the Phylum Chordata.

3. All of the previous animals we have studied in Chapters 4-11 are ____________________, animals without a spinal cord or vertebrae (backbones) such as starfish, insects, oysters, etc.

4. Name the three classes of fish living today. (See pages 99, 117, and 125.)

   a.

   b.

   c.

5. All three classes of fish are members of the Phylum ____________________.

6. ______________________ fish are the largest class of fish living today.

7. List nine modern types of bony fish found in dinosaur rock layers: (See pages 100-110.)

   a.

   b.

   c.

   d.

   e.

   f.

   g.

   h.

   i.

8. Give the two *subclasses* of bony fish living today and examples of fish within them. Note: both subclasses have been found with dinosaurs.

   a.

   b.

9. The author predicted he would find modern types of fish in dinosaur layers if evolution is not true. Is this what he found?

Purpose of Chapter: To test the author's idea that if evolution was *not* true, modern-appearing cartilaginous fish should be found with dinosaurs.

Answer the Discussion Questions. Compare your answers to the next page:

1. How do cartilaginous fish, such as sharks and rays, differ from bony fish?

2. What do you think the author predicted he would find regarding cartilaginous fish such as sharks and rays if evolution was *not* true?

3. From what you have seen in museums or in television documentaries, did modern types of sharks and rays live at the same time as dinosaurs?

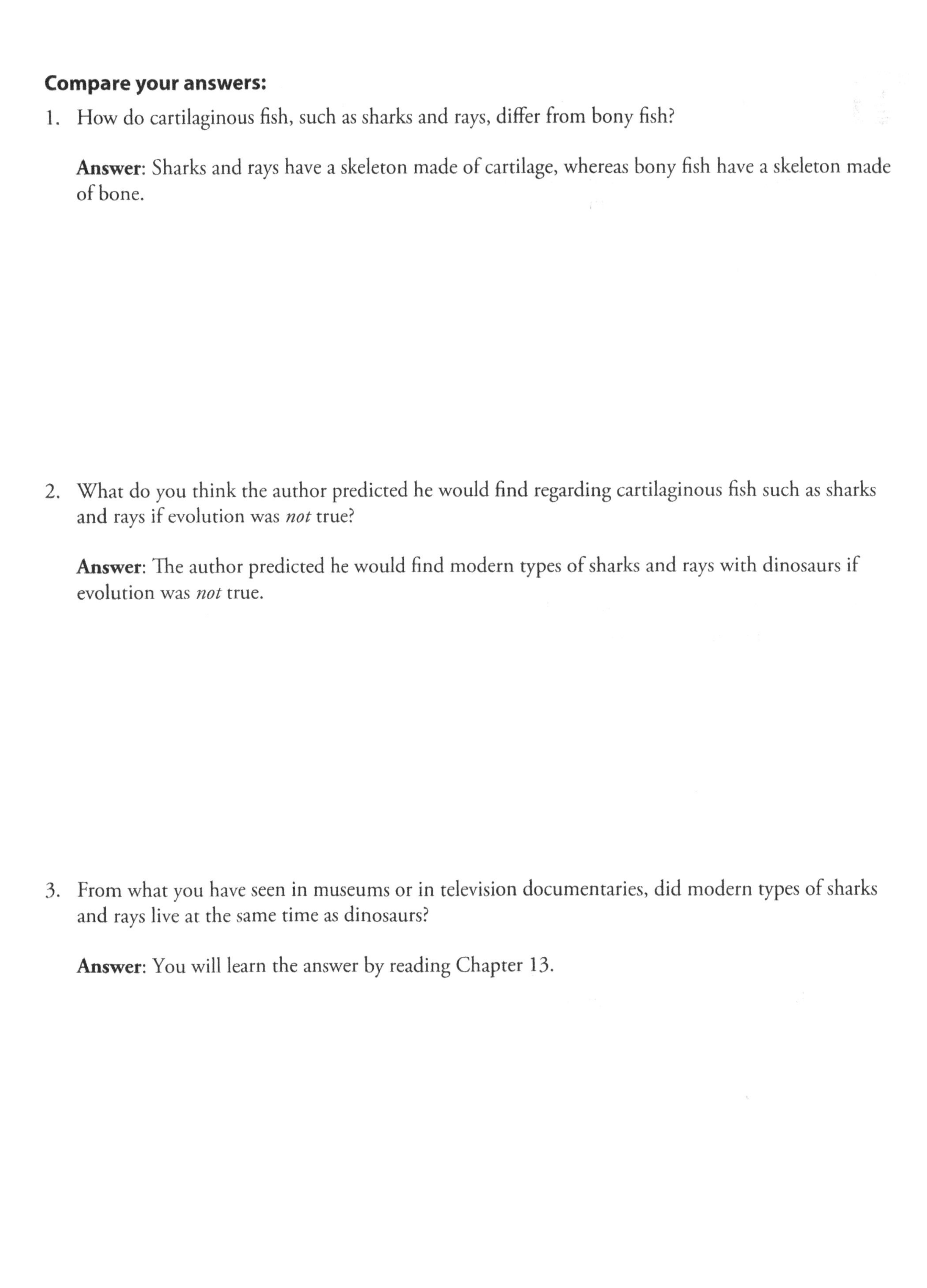

Compare your answers:

1. How do cartilaginous fish, such as sharks and rays, differ from bony fish?

   Answer: Sharks and rays have a skeleton made of cartilage, whereas bony fish have a skeleton made of bone.

2. What do you think the author predicted he would find regarding cartilaginous fish such as sharks and rays if evolution was *not* true?

   Answer: The author predicted he would find modern types of sharks and rays with dinosaurs if evolution was *not* true.

3. From what you have seen in museums or in television documentaries, did modern types of sharks and rays live at the same time as dinosaurs?

   Answer: You will learn the answer by reading Chapter 13.

1. ______________________ fish, such as sharks and rays, have skeletons made of cartilage instead of bone.

2. Name the three classes of fish living today. (See pages 99, 117, and 125.)

   a.

   b.

   c.

3. All classes of fish belong to the Phylum _________________ because they have a spinal cord.

4. A shark was found in Solnhofen, Germany (along with the dinosaur *Compsognathus* and the famous toothed bird *Archaeopteryx*), that looked ____________________ to a modern angel shark.

5. A second shark was found in Solnhofen, Germany, which looked similar to a modern _______________ _______________ shark.

6. A ray was found in Solnhofen, Germany, which looked similar to a modern __________________ ray.

7. A shark tooth was found in dinosaur rock layers in _________ _________________ which looked similar to a modern goblin shark. Even though it looked similar to the modern form, the scientist who discovered the fossil gave it a new genus and species name.

8. Shark teeth were found buried in dinosaur rock layers at Dinosaur Provincial Park in ___________, Canada, which looked similar to modern shark teeth.

9. The author predicted he would find modern types of cartilaginous fish in dinosaur layers if evolution was not true. Is this what he found?

10. Fill in this chart:

Rock Layer, Location	Fossil(s) that Looks Similar to Modern Fish
Jurassic, Solnhofen, Germany	
Cretaceous, New Mexico, USA	
Cretaceous, Alberta, Canada	

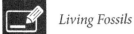

Purpose of Chapter: To test the author's idea that if evolution was *not* true, modern-appearing jawless fish should be found with dinosaurs.

Answer the Discussion Questions. Compare your answers to the next page:

1. There are two types of jawless fish living today. Can you name them?

2. Evolution scientists believe jawless fish are the most primitive fish and that they evolved into other fish. If this is true, why are jawless fish living today?

Compare your answers:

1. There are two types of jawless fish living today. Can you name them?

   Answer: The answer is in Chapter 14.

2. Evolution scientists believe jawless fish are the most primitive fish and that they evolved into other fish. If this is true, why are jawless fish living today?

   Answer: If animals were eliminated over time through the principle of the survival of the fittest, then one would expect the most primitive classes of fish (such as the jawless fish) would not be alive today, but they are. This seems contradictory to the theory of evolution.

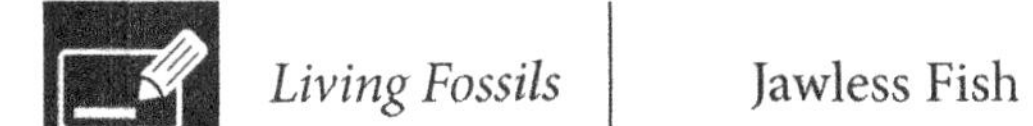

Living Fossils	Jawless Fish	Day 135	Chapter 14 Worksheet	Name

1. Give the two types of jawless fish living today (page 125).
   a.

   b.

2. The author predicted finding modern types of jawless fish in dinosaur rock layers if evolution is not true. Is this what he found?

3. According to Dr. John Long, a proponent of the theory of evolution and author of *The Rise of Fishes*, ____________________ fish living today (lampreys and the hagfish) have not changed much since the dinosaur era.

4. List the three classes of fish living today. Note: all three classes have been found in dinosaur rock layers and appear basically unchanged.
   a.

   b.

   c.

5. According to the theory of evolution, ________________________ fish are the oldest fish group. Even so, this "ancient fish class" is still alive today and this contradicts the evolutionary idea of change over time.

Purpose of Chapter: To test the author's idea that if evolution was *not* true, modern-appearing amphibians should be found with dinosaurs.

Answer the Discussion Questions. Compare your answers to the next page:

1. There are two major types of amphibians living today. Do you know what the two major groups are?

2. The author predicted he would find modern amphibians with dinosaurs if evolution was not true. Have you ever seen a museum display or television documentary suggesting that modern-appearing frogs and salamanders lived with dinosaurs?

3. Do you believe that *T. rex* lived alongside frogs and salamanders?

4. If all the animals were created at one time, such as dinosaurs, frogs, bluegill, and dogs, as the biblical account of Creation suggests, would you expect to find *modern* frogs and salamanders in dinosaur rock layers?

Compare your answers:

1. There are two major types of amphibians living today. Do you know what the two major groups are?

   Answer: Frogs/toads and salamanders.

2. The author predicted he would find modern amphibians with dinosaurs if evolution was not true. Have you ever seen a museum display or television documentary suggesting that modern-appearing frogs and salamanders lived with dinosaurs?

   Answer: Answers may vary.

3. Do you believe that *T. rex* lived alongside frogs and salamanders?

   Answer: The answer will be found when you read chapter 15.

4. If all the animals were created at one time, such as dinosaurs, frogs, bluegill, and dogs, as the biblical account of Creation suggests, would you expect to find *modern* frogs and salamanders in dinosaur rock layers?

   Answer: Yes. If all the animals were created at the same time, you would expect to find some fossils of modern amphibians and other modern animals with dinosaurs.

Living Fossils	Amphibians	Day 138	Chapter 15 Worksheet	Name

1. Amphibians are vertebrates and are members of the Phylum ____________________.

2. List the five major groups of this phylum (see page 98):

   a.

   b.

   c.

   d.

   e.

3. There are only two *major* amphibian groups living today,________________ /________________ and ________________________, and both groups were also alive during the time of the dinosaurs. (The third group of amphibians called legless salamanders is a *minor* amphibian group and is not addressed in this book.)

4. A fossil __________________________ was found with 29 *Iguanodon* dinosaurs in a coal mine in Bernissart, Belgium.

5. Another dinosaur-era salamander called __________________________, found in dinosaur rock layers, was like a modern salamander.

6. Fossil ________________ have been found in Jurassic dinosaur rock layers.

7. The author predicted that if evolution was *not* true, modern appearing amphibians should be found in dinosaur rock layers. Is this what he found?

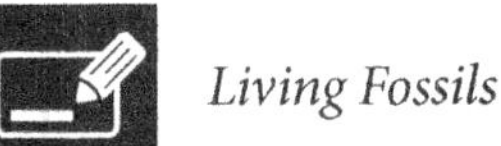

Living Fossils	Crocodilians	Day 140	Chapter 16 Discussion	Name

Purpose of Chapter: To test the author's idea that if evolution was *not* true, modern-appearing crocodilians should be found with dinosaurs.

Answer the Discussion Questions. Compare your answers to the next page:

1. There are five major groups of animals that are vertebrates (animals with backbones). Excluding fish and amphibians, what are the other three major groups of vertebrates?

2. Which reptile lives in the Florida Everglades: alligators or crocodiles?

3. Which lives in Africa and jumps out of the water attacking large mammals: alligators or crocodiles?

Compare your answers:

1. There are five major groups of animals that are vertebrates (animals with backbones). Excluding fish and amphibians, what are the other three major groups of vertebrates?

   Answer: Reptiles, birds, and mammals.

2. Which reptile lives in the Florida Everglades: alligators or crocodiles?

   Answer: Alligators.

3. Which lives in Africa and jumps out of the water attacking large mammals: alligators or crocodiles?

   Answer: Crocodiles.

1. List the three types of crocodilians living today:

   a.

   b.

   c.

2. Crocodilians are vertebrates, members of the Phylum ________________________.

3. Alligators today live along the Gulf Coast of the United States, and they have broader __________________________ than crocodiles.

4. A fossil crocodile skull was found at Dinosaur Provincial Park in ____________________, Canada.

5. Alligator and crocodile skulls were found in the dinosaur rock layers at Dinosaur Provincial Park. Even though these fossil skulls look ____________________ to modern varieties, they were assigned different genus and species names.

6. _____________________________ live today in Bangladesh, India, Pakistan, and Burma.

7. A gavial-like reptile was found in dinosaur rock layers in ______________________, and it looked similar to modern gavials.

8. List all three types of crocodilians living today that were also alive during the time of the dinosaurs and looked similar to modern forms, but they were assigned new genus and species names.

   a.

   b.

   c.

9. The author predicted he would find modern-appearing crocodilians in dinosaur rock layers, if evolution was *not* true. Is this what he found?

Living Fossils	Snakes	Day 143	Chapter 17 Discussion	Name

Purpose of Chapter: To test the author's idea that if evolution was *not* true, modern-appearing snakes should be found with dinosaurs.

Answer the Discussion Questions. Compare your answers to the next page:

1. Have you ever gotten the impression from science museums or from watching educational television programs that boa constrictors lived with *T. rex*?

2. When the author began his experiment, he predicted that if evolution is *not* true, then he would find modern types of snakes with dinosaurs. What do you predict he will find based on what you have learned so far?

Compare your answers:

1. Have you ever gotten the impression from science museums or from watching educational television programs that boa constrictors lived with *T. rex*?

   Answer: Answers will vary.

2. When the author began his experiment, he predicted that if evolution is *not* true, then he would find modern types of snakes with dinosaurs. What do you predict he will find based on what you have learned so far?

   Answer: Answers will vary.

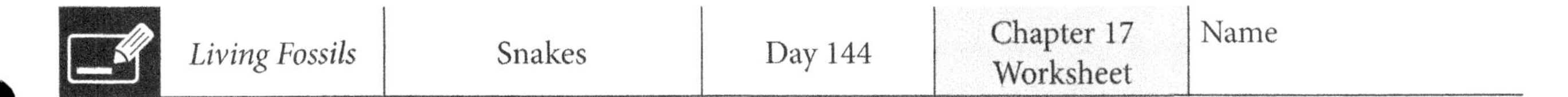

1. What phylum includes snakes?

2. List the seven major animal phyla living today. Be aware that all seven have been found with dinosaurs and appear relatively unchanged. (See pages 29, 45, 71, 89, 93, and 99.)

   a.

   b.

   c.

   d.

   e.

   f.

   g.

3. A fossil __________ ________________________ was found at Hell Creek, Montana, along with a *T. rex* and a *Triceratops* dinosaur.

4. The fossil snake found at Hell Creek looked like a ____________________ _______________ ____________________________.

5. Fossil ______________________ have been found in dinosaur rock layers in at least nine countries so far.

Purpose of Chapter: To test the author's idea that if evolution was *not* true, modern-appearing lizards should be found with dinosaurs.

Answer the Discussion Questions. Compare your answers to the next page:

1. Have you ever gotten the impression from watching educational television programs that modern-appearing lizards lived with dinosaurs?

2. When the author began his experiment, he predicted that if evolution is *not* true, then he would find modern lizards with dinosaurs. What do you think he found based on the other animal groups you have studied so far?

Compare your answers:

1. Have you ever gotten the impression from watching educational television programs that modern-appearing lizards lived with dinosaurs?

   Answer: Answers will vary.

2. When the author began his experiment, he predicted that if evolution is *not* true, then he would find modern lizards with dinosaurs. What do you think he found based on the other animal groups you have studied so far?

   Answer: Answers will vary.

1. Lizards have backbones called vertebrae, have a spinal cord, and are members of the Phylum ____________________________.

2. A fossil iguana-like lizard was found in dinosaur rock layers in __________________ and is now housed at the Carnegie Museum.

3. Ground __________________________ today live all over the world.

4. A fossil ground lizard was found in Germany that looked like a modern _______________________ lizard.

5. A fossil lizard was found in Germany in the same rock layers as the dinosaur *Compsognathus* and the toothed bird *Archaeopteryx*. This fossil lizard looks very similar to a ______________________________ tuatara lizard.

6. Gliding lizards have elongated _________________________ bones that support their membranous "wings."

7. A fossil gliding lizard was found in _______________________ dinosaur rock layers in the United States.

8. Fill in the chart (page 140).

Rock Layer, Location	Fossil Looks Similar to This Modern Lizard
Cretaceous, Utah, USA	
Jurassic, Solnhofen, Germany	
Jurassic, Eichstatt, Germany	
Triassic, New Jersey, USA	

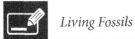

Purpose of Chapter: To test the author's idea that if evolution was *not* true, modern-appearing turtles should be found with dinosaurs.

Answer the Discussion Questions. Compare your answers to the next page:

1. There are five major groups of vertebrates (animals with backbones). Excluding fish, what are the other four major groups?

2. This chapter deals with the last group of reptiles living today, turtles. From visiting museums and watching educational television, is it your impression that modern types of turtles (such as box turtles and soft-shelled turtles) lived with the dinosaurs?

3. What do you think the author predicted he would find regarding turtles if evolution is *not* true?

Compare your answers:

1. There are five major groups of vertebrates (animals with backbones). Excluding fish, what are the other four major groups?

   Answer: Amphibians, reptiles, birds, and mammals.

2. This chapter deals with the last group of reptiles living today, turtles. From visiting museums and watching educational television, is it your impression that modern types of turtles (such as box turtles and soft-shelled turtles) lived with the dinosaurs?

   Answer: The answer will be found in chapter 19.

3. What do you think the author predicted he would find regarding turtles if evolution is *not* true?

   Answer: The author predicted that if evolution is *not* true, he would find modern types of turtles in dinosaur rock layers.

1. Turtles are vertebrates, have a ______________________ ______________________, and belong to the Phylum Chordata.

2. List the five groups of vertebrates living today:
   a.
   b.
   c.
   d.
   e.

3. List the four groups of reptiles living today:
   a.
   b.
   c.
   d.

4. The author predicted that if evolution was not true, he would find modern-appearing turtles in ____________________ rock layers. (See page 28.)

5. A fossilized box turtle was found at Hell Creek, Montana, along with dinosaurs in ______________________ rock layers.

6. A fossil ______________________ was found in Jurassic rock layers in Germany in the same rock layers as the dinosaur *Compsognathus* and the toothed bird *Archaeopteryx*.

7. There are two sub-orders of turtles living today, the ____________-____________ turtles (Pleurodira) and the ____________-____________ turtles (Cryptodira). Both sub-orders of turtles living today have also been found in dinosaur rock layers.

8. List all seven animal phyla living today. Note: they appear basically unchanged (see pages 29, 45, 71, 89, 93, and 99).

   a.

   b.

   c.

   d.

   e.

   f.

   g.

9. If animals changed over time, as described in the theory of evolution, one would not expect to find ____________________ types of turtles in dinosaur rock layers.

10. If all the animals were created at one time, such as described in the biblical account of creation, one would expect to find modern types of turtles and other modern types of animals in ______________________ rock layers.

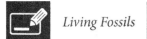

Purpose of Chapter: To test the author's idea that if evolution was *not* true, modern-appearing birds should be found with dinosaurs.

Answer the Discussion Questions. Compare your answers to the next page:

1. There are five major groups that comprise vertebrates (animals with backbones). Fish were the first group. What are the other four major groups of vertebrates?

2. This chapter deals with birds. From what you have learned from museums and educational television, did modern birds, such as parrots and flamingos, live with the dinosaurs?

3. What do you think the author predicted about modern birds if evolution was *not* true?

Compare your answers:

1. There are five major groups that comprise vertebrates (animals with backbones). Fish were the first group. What are the other four major groups of vertebrates?

   Answer: Amphibians, reptiles, birds, and mammals.

2. This chapter deals with birds. From what you have learned from museums and educational television, did modern birds, such as parrots and flamingos, live with the dinosaurs?

   Answer: The answer will be found in chapter 20.

3. What do you think the author predicted about modern birds if evolution was *not* true?

   Answer: The author predicted that if evolution was *not* true, he would find modern birds in dinosaur rock layers.

1. Birds belong to the Phylum ______________________ because they have a spinal cord and vertebrae.

2. List the five groups of vertebrates living today. Be aware that all five of these vertebrate groups were also alive during the time of the dinosaurs. (See pages 99, 129, 137, 161 and 169.)
   a.
   b.
   c.
   d.
   e.

3. The author predicted he would find modern-appearing birds in dinosaur rock layers if ______________________ was not true. (See page 28.)

4. List ten modern bird types found in dinosaur rock layers:
   a.
   b.
   c.
   d.
   e.
   f.
   g.
   h.
   i.
   j.

5. Name three birds living with dinosaurs that went extinct. (See pages 180-181 of Volume I, *Evolution: The Grand Experiment, The Quest for an Answer.*)

   a.

   b.

   c.

6. Based on ______________________ evidence, most or all of the major modern bird groups were present in the Cretaceous.

7. *All* seven animal phyla living today have been found with ____________________ and they appear basically unchanged.

Purpose of Chapter: To test the author's idea that if evolution was *not* true, modern-appearing mammals should be found with dinosaurs.

Answer the Discussion Questions. Compare your answers to the next page:

1. Did mammals live during the time of the dinosaurs?

2. What did the author predict he would find about mammals if evolution was not true?

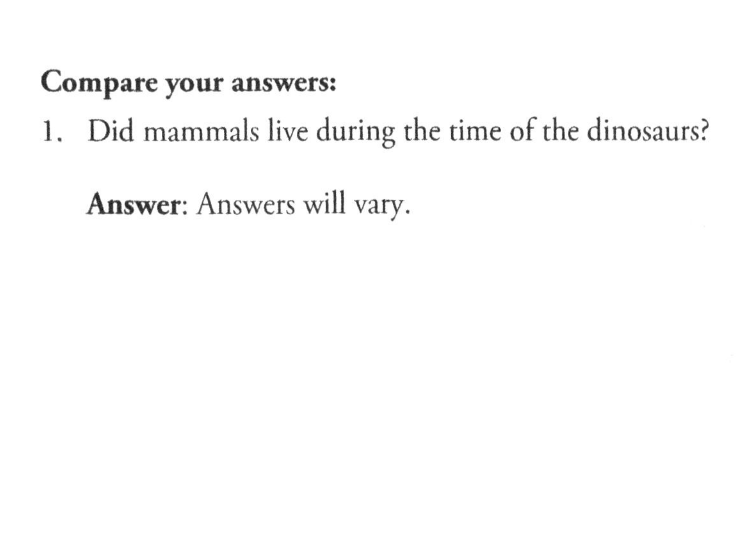

Compare your answers:

1. Did mammals live during the time of the dinosaurs?

   Answer: Answers will vary.

2. What did the author predict he would find about mammals if evolution was *not* true?

   Answer: The author predicted he would find *modern* types of mammals in dinosaur rock layers if evolution was *not* true.

1. Mammals are vertebrates, have a spinal cord, and belong to the Phylum ______________________.
2. List the three subclasses of mammals living today (page 169):

   a.

   b.

   c.

3. "Negative evidence," or the ______________________ of things, is a tricky issue to deal with. In other words, if a fossil is missing from a particular dinosaur rock layer, it does not necessarily mean that a particular animal did not live at that time.

4. What year were mammal bones first discovered in dinosaur rock layers? Compare this to the year that Darwin wrote *The Origin of Species* in 1859.

   a.

   b.

5. Nearly ______________ different genus groups of mammals have now been found in dinosaur layers. Many of these 300 genus groups have more than one species.

6. Mammals have been found in all three dinosaur rock layers. List the rock layers:

   a.

   b.

   c.

7. Explain why the term "Age of the Reptiles" when referring to dinosaur times is a misnomer. Write out the second sentence of Dr. Luo's quote at the bottom of page 172.

   a.

   b.

8. For over 100 years, it was taught that dinosaur-era mammals were very small, about the size of a shrew or mouse. Recently a 30-pound mammal (that looked like a ____________________ ____________________) was found in dinosaur rock layers.

9. Give the species names and weight of the two larger mammals found with dinosaurs.

   a. (Possum-like)-

   b. (Size of a collie)-

1. According to Dr. Luo, how many complete mammal skeletons have been found in dinosaur rock layers? How many species do these complete skeletons represent?

   a.

   b.

2. The author only saw ______________ of these nearly 100 *complete* mammal skeletons found in dinosaur rock layers at the 60 museums he visited.

3. A mammal was found in dinosaur rock layers in Australia (*Ausktribosphenos nyktos*) that looked similar to a modern ______________________.

4. The museum reconstruction of this mammal (*Ausktribosphenos nyktos*) looked similar to a modern ______________________.

5. A mammal (*Gobiconodon*) was found in dinosaur rock layers that looked similar to a ringtail ______________________.

6. One museum reconstruction of *Gobiconodon* looked eerily similar to a modern ringtail ______________________.

7. All three ______________________ of mammals living today, placental, marsupial, and egg-laying Monotreme, were also living with the dinosaurs.

8. List four other mammals found in dinosaur layers (page 182):

   a.

   b.

   c.

   d.

9. According to Dr. Dawson, opossums, duck-billed platypus, echidnas, and even shrews had ________________counterparts.

Purpose of Chapter: To test the author's idea that if evolution was *not* true, modern-appearing cone-bearing plants should be found with dinosaurs.

Answer the Discussion Questions. Compare your answers to the next page:

1. Name some trees that have cones.

2. The California redwood trees are beautiful cone trees, some as large as 300 feet tall and some so wide you could drive a car through the trunk. Have you ever seen pictures of redwoods or seen them while on vacation in California? There is a picture of one on page 186 of the book.

3. From what you have learned on public television or in museums that support evolution, did dinosaurs live at the same time as redwood trees?

4. What does the theory of evolution teach about the types of trees that lived during dinosaur times?

5. What does the theory of creation teach about the types of trees that lived during dinosaur times?

6. What do you think the author predicted about cone trees?

Compare your answers:

1. Name some trees that have cones.

   Answer: Pine trees, redwoods, sequoias, cypress, cycad, ginkgo, etc.

2. The California redwood trees are beautiful cone trees, some as large as 300 feet tall and some so wide you could drive a car through the trunk. Have you ever seen pictures of redwoods or seen them while on vacation in California? There is a picture of one on page 186 of the book.

   Answer: Answers will vary.

3. From what you have learned on public television or in museums that support evolution, did dinosaurs live at the same time as redwood trees?

   Answer: Answers will vary.

4. What does the theory of evolution teach about the types of trees that lived during dinosaur times?

   Answer: The theory of evolution says that plants changed over time and that in general modern plant species were not alive during the time of the dinosaurs.

5. What does the theory of creation teach about the types of trees that lived during dinosaur times?

   Answer: The theory of creation suggests that all of the animals and plants that have ever lived on earth, including modern plants and modern animals, were created at one time and lived side by side.

6. What do you think the author predicted about cone trees?

   Answer: The author predicted that if evolution was not true, then he would find fossils of modern types of cone trees, such as sequoias, pine trees, cypress trees, cycads, etc., in dinosaur rock layers.

1. There are only seven *major* animal phylum living today, and all have been found with dinosaurs. List these seven phyla and give at least one example from each phylum (page 233).

   a.

   b.

   c.

   d.

   e.

   f.

   g.

2. The word phylum is used for animals but the word ____________________ is used for plants. (See page 16.)

3. Three of the seven major plant divisions living today are cone bearing plants. List them: (See pages 183, 240.)

   a.

   b.

   c.

4. Give the names of the three divisions of cone-bearing plants living today:

   a. Cone trees are Division-

   b. Cycads are Division-

   c. Gingko trees are Division-

5. All three divisions of cone-bearing plants, grouped together, are called________________________. (See page 185.)

6. What is the genus name for California redwood trees? (See page 186.)

7. List the conifers (Division Coniferae) that have been found in dinosaur layers which look similar to modern forms:

   a.

   b.

   c.

   d.

   e.

8. Name two dinosaur-era cycad trees (Division Cycadophyta).

   a.

   b.

9. ____________________ are the third major type of cone-bearing plants.

10. All three divisions of cone bearing plants living today (Division Coniferae, Division Cycadophyta, Division Ginkgophyta) have also been found in ________________________ rock layers.

11. The author predicted that if evolution was not true he would find modern-appearing cone trees in dinosaur rock layers. Is this what he found?

Purpose of Chapter: To test the author's idea that if evolution was *not* true, modern-appearing spore-forming plants should be found with dinosaurs.

Answer the Discussion Questions. Compare your answers to the next page:

1. Cone-bearing plants reproduce with cones. Flowering plants reproduce with flowers. How do you think spore-forming plants reproduce?

2. Horsetail plants, ferns and moss reproduce with spores. Have you ever seen a horsetail plant? What do they look like?

3. What do you think the author predicted he would find regarding spore-forming plants if evolution was *not* true?

Compare your answers:

1. Cone-bearing plants reproduce with cones. Flowering plants reproduce with flowers. How do you think spore-forming plants reproduce?

   Answer: With spores.

2. Horsetail plants, ferns and moss reproduce with spores. Have you ever seen a horsetail plant? What do they look like?

   Answer: There is a picture of one in the *Living Fossils* book on page 205.

3. What do you think the author predicted he would find regarding spore-forming plants if evolution was *not* true?

   Answer: The author predicted that if evolution was not true, he would find modern-appearing spore-forming plants, such as ferns, horsetails, and moss, in dinosaur rock layers.

1. (Review question) Give the three divisions of plants living today that reproduce with cones and give at least one example from each division. (See pages 183, 240.)

   a.

   b.

   c.

2. List the three divisions of plants living today that reproduce with spores and give at least one example from each division. (See pages 197, 240.)

   a.

   b.

   c.

3. All three divisions of spore-forming plants living today were also alive at the time of the ____________________________ and appear nearly the same.

4. Give the genus name for peat moss that is used in gardens today. This same genus of plant was found in dinosaur rock layers.

Purpose of Chapter: To test the author's idea that if evolution was *not* true, modern-appearing flowering plants should be found with dinosaurs.

Answer the Discussion Questions. Compare your answers to the next page:

1. There are seven major divisions of plants living today:

   Three divisions are ______________________________

   Three divisions are ______________________________

   One division is the ______________________________ which are by far the biggest plant division.

   Name some *plants* that have flowers.

2. Name some *trees* that have flowers.

3. What do you think the author predicted he would find regarding flowers if evolution was *not* true?

4. Have you ever seen a documentary or science museum display suggesting that modern flowering plants lived with dinosaurs?

Compare your answers:

1. There are seven major divisions of plants living today:

   Three divisions are cone-bearing plants.

   Three divisions are spore-bearing plants.

   One division is the flowering plants. Flowering plants, by far, are the biggest plant division.

   Name some *plants* that have flowers.

   Answer: Roses, tomatoes, daisies, etc.

2. Name some *trees* that have flowers.

   Answer: Apple, dogwood, maple, etc.

3. What do you think the author predicted he would find regarding flowers if evolution was *not* true?

   Answer: He predicted he would find modern types of flowering plants in dinosaur rock layers if evolution is not true.

4. Have you ever seen a documentary or science museum display suggesting that modern flowering plants lived with dinosaurs?

   Answer: Answers will vary.

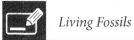

1. Flowering plants make up one of the seven major plant divisions. To which plant division (name of division) do flowering plants belong?

2. List the seven major plant divisions living today and give an example from each division (page 240).

   a.

   b.

   c.

   d.

   e.

   f.

   g.

3. ______________________ plants, by far, are the most common type of plants living on the earth.

4. Evolution scientists, museums, and textbooks have made obtuse suggestions that modern flowering plants were not alive during the time of the ________________. They have implied that only non-flowering plants (ferns, cycads, and mosses) were alive then.

5. Give the names of the three rock layers that have dinosaurs in them. (See page 10.)

   a.

   b.

   c.

6. List 15 common flowering plants found in dinosaur rock layers (and be able to write out this list from memory). (See pages 213-225 and page 240.)

   a.

   b.

   c.

   d.

   e.

   f.

   g.

   h.

   i.

   j.

   k.

   l.

   m.

   n.

   o.

7. The author predicted that he would find ______________________ plants in dinosaur rock layers if evolution was *not* true. He found fossilized modern-appearing examples of all seven major plant divisions living today in the ________________________ rock layers thus giving him evidence that evolution is *not* true.

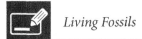

Purpose of Chapter: To test the author's idea that if evolution was *not* true, modern-appearing animals and plants should be found with dinosaurs.

Answer the Discussion Questions. Compare your answers to the next page:

1. The author predicted that if evolution was not true, he would find modern plants and animals in the dinosaur rock layers. He did not find all of them, such as horses or dogs, but found many of them. Does the absence of a fossil mean that a particular plant or animal did not live with dinosaurs?

Compare your answers:

1. The author predicted that if evolution was not true, he would find modern plants and animals in the dinosaur rock layers. He did not find all of them, such as horses or dogs, but found many of them. Does the absence of a fossil mean that a particular plant or animal did not live with dinosaurs?

   Answer: Remember the concept of negative evidence. Specifically, the absence of a particular plant or animal in any particular rock layer does not necessarily mean the plant or animal did not live at that time. Read the quote on the bottom of page 170 before proceeding to read Chapter 25.

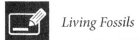

1. When the author ignored the "subjective, scientist-generated" genus and species names and simply compared the fossils found in dinosaur rock layers to modern forms, he saw a lack of significant ___________________________ in all of the major animal phyla and all of the major plant divisions.

2. After comparing the fossils found in dinosaur rock layers to modern forms, the author concluded that animals and plants have not ____________________ (evolved) dramatically over time, but simply some animals and plants have gone ___________________ (such as dinosaurs, pterosaurs and the like), while others have remained relatively unchanged.

3. The author only had access to 20,000 of the 200,000,000 fossils collected by museums or less than .01% of the worldwide collections of fossils.[1] Despite this severe limitation, he found dinosaur-era fossil examples of all seven *major* animal phyla living today and all seven *major* plant divisions living today in the_______________________________ rock layer which looked similar to modern forms.

4. The author, despite his diligence of traveling to 60 museums, saw only three (one was not exhibited) of the nearly 100 complete mammal skeletons found in dinosaur rock layers, and only a handful of the nearly 300 genera of the collected dinosaur-era mammal fossils. Despite this severe limitation, he found examples of all three subclasses of mammals living today in ________________________ rock layers which look similar to modern mammals.

5. Just because a particular animal is not found in a particular rock layer does not mean that it did not exist at that time. This "absence of things" is called ______________________ ______________________. (See interview at the bottom of page 170.)

1 As of 2008, the fossil count is nearly one billion. See footnote 1 in Appendix A of *Evolution: The Grand Experiment.*

6. In addition to living fossils, there are seven other major problems with the current theories explaining the origin of the universe and the origin of life in natural terms. Using the subtitles from pages 241-243, list and explain these seven major problems.

   a.

   b.

   c.

   d.

   e.

   f.

   g.

7. "The manner in which fossils are ______________________ may aid in the perception that animals (or plants) changed dramatically over time or may aid in the perception that they did not change over time." (See Appendix A, page 244.)

Chapter Tests

for Use with

Evolution: The Grand Experiment

The Grand Experiment	Chapter 1 Test	The Origin of Life: Two Opposing Views	Total score: ____of 100	Name
Concepts & Comprehension				

1. What are the two opposing views on the origin of life?

   a.

   b.

2. Name *one* of the three best *fossil* evidences *for* evolution, cited by scientists who support the theory.

3. Name one of the three major scientific developments that has occurred since Darwin first published his theory of evolution in 1859.

4. Name *one* of the four best evidences *against* evolution, cited by scientists who oppose the theory.

5. Michelangelo's famous painting from the ceiling of the Sistine Chapel depicts what?

6. According to a Gallup poll taken in 2012, many Americans, 46 percent, believe God created man in the last ________ years.

7. What percentage of Americans believe evolution did occur, but that God guided the process?

8. *True or False*: Only 15 percent of Americans believe in evolution, that humans evolved from apes, and God had no part in the process.

9. *True or False*: The majority of Americans believe creationism should *not* be taught in *public* schools.

10. Name one of the fears educators have in teaching two opposing theories about the origin of life.

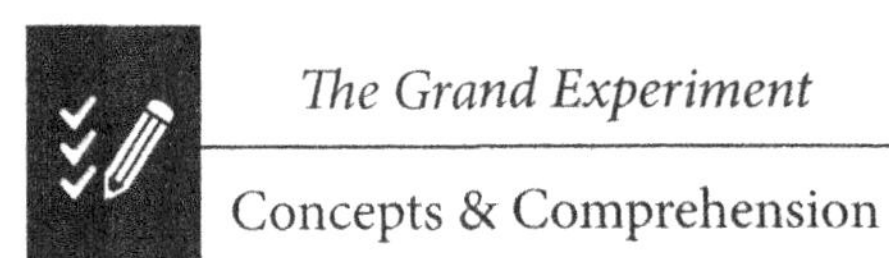

The Grand Experiment / Concepts & Comprehension	Chapter 2 Test	Evolution's False Start: Spontaneous Generation	Total score: ____of 100	Name

1. Which scientist thought he proved that life evolved from dirty underwear?

2–4. Name the three "proofs" of spontaneous generation (that subsequently were disproved).

a.

b.

c.

5. In what year did Italian physician and scientist Dr. Francesco Redi disprove the maggots from meat experiment?

6. Pond scum that forms in a bottle of boiled pond water is actually a collection of ______________.

7. Who disproved the spontaneous generation of bacteria?

8. How many years was spontaneous generation believed by the majority of scientists until it was finally disproved?

9. In what year was spontaneous generation finally disproved?

10. What theory eventually replaced the theory of spontaneous generation and is the basis for the modern theory of how life began naturally?

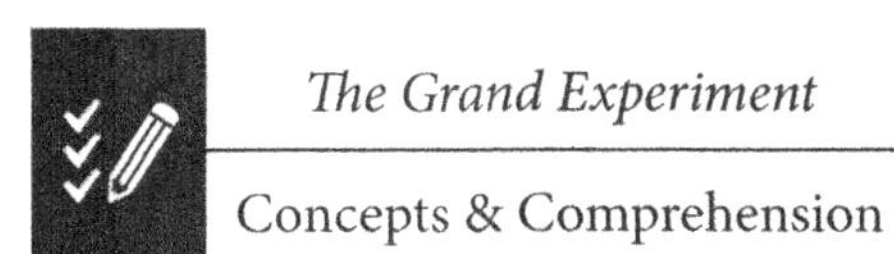

The Grand Experiment / Concepts & Comprehension	Chapter 3 Test	Darwin's False Mechanism for Evolution	Total score: ____of 100	Name

1. In what year did Darwin publish his first book on evolution?

2. Define the theory of acquired characteristics.

3. Name four examples of acquired characteristics that ultimately were shown to be wrong.
   a.

   b.

   c.

   d.

4. *True or False:* Darwin was proved wrong on the law of disuse (or the shedding of body parts).

5. *True or False:* Acquired characteristics are a valid mechanism by which animals change over time.

6. Write out Darwin's quote which demonstrates he believed in acquired characteristics at the end of his lifetime.

7. What is another name for the law of acquired characteristics? The law of __________ and ________________.

8. Which scientist disproved the law of disuse?

9. How was the law of disuse disproved?

10. If you place your left arm in a sling for the rest of your life and let the muscles of your left arm become small or weak, will your future offspring then be born with an arm that is weak or small? Why or why not?

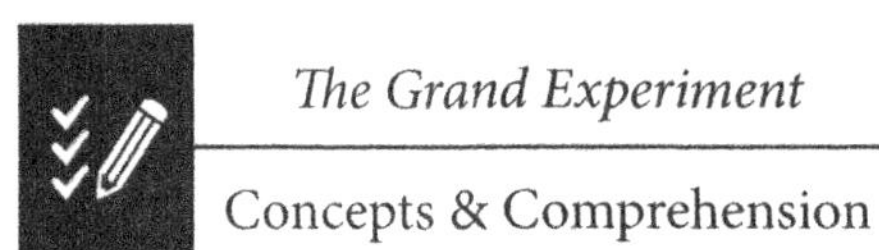

The Grand Experiment / Concepts & Comprehension	Chapter 4 Test	Natural Selection and Chance Mutations	Total score: ____of 100	Name

1. What animal order is composed of whales, dolphins, and porpoises?

2. *True or False:* Natural selection or artificial breeding alone can add completely new body parts to an animal, such as a fin or a feather or a cardiovascular system.

3. What is possibly the largest animal to have ever lived on earth?

4. Charles Darwin reasoned that whales could have evolved from what land mammal?

5. Today scientists recognize that an animal could only evolve by accidental ______________ in the DNA of reproductive cells.

6. List three animals that evolution scientists believe may have evolved into whales.

   a.

   b.

   c.

7. What new trait did Dr. Morgan observe which made him realize that accidental mutations accounted for new traits?

8. Name the 4 letters that make up DNA. _____ _____ _____ _____

9. *True or False*: Mutations of DNA can be dangerous to an animal's health.

10. Name five ways a hyena would have to bodily change, by chance mutations, in order for it to evolve into a whale.

    a.

    b.

    c.

    d.

    e.

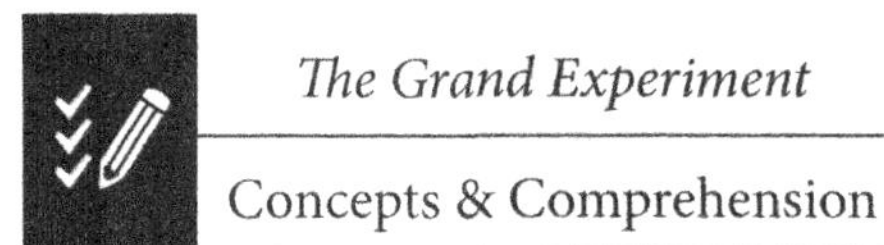

The Grand Experiment / Concepts & Comprehension	Chapter 5 Test	Similarities: A Basic Proof of Evolution?	Total score: ____of 100	Name

1. According to evolution scientists, is a placental mole more closely related to a pouched mole or a whale?

2. According to evolution scientists, is a placental mouse more closely related to a pouched mouse or a horse?

3. Write out the definition of "convergent evolution."

4. Why do scientists who oppose evolution object to using similarities as "proof" of evolution?

5. What two animals do evolution scientists believe sea lions may have evolved from?
   a.

   b.

6. What two animals do evolution scientists believe seals may have evolved from?
   a.

   b.

7. Explain why, according to evolution scientists, similarities of the seal and sea lion do not equate to evolution.

8–10. Name three animals that have wings but are unrelated.

a.

b.

c.

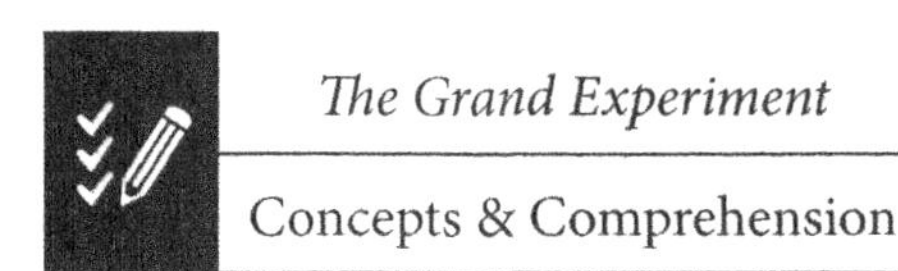

The Grand Experiment / Concepts & Comprehension	Chapter 6 Test	The Fossil Record and Darwin's Prediction	Total score: ____of 100	Name

1. *True or False*: Darwin recognized that the fossil record did not match what his theory predicted.

2. *True or False*: Since Darwin's time, millions of (difficult-to-fossilize) soft-bodied plants and animals have been found as fossils.

3. What percentage of living land animal *families* (*including birds*) have been found as fossils?

4. How many fossil fish have been *collected* by museums?

5. *True or False*: If evolution is true and if the fossil record is nearly complete, the fossil record should show one animal slowly changing into a completely different type of animal over time.

6. How many fossil bats have been *collected* by museums?

7. How many fossil insects have been *collected* by museums?

8. How many fossils (total number of all types of fossils) have been *collected* by museums?

9. How many dinosaurs have been *collected* by museums?

10. How many fossil birds have been *collected* by museums?

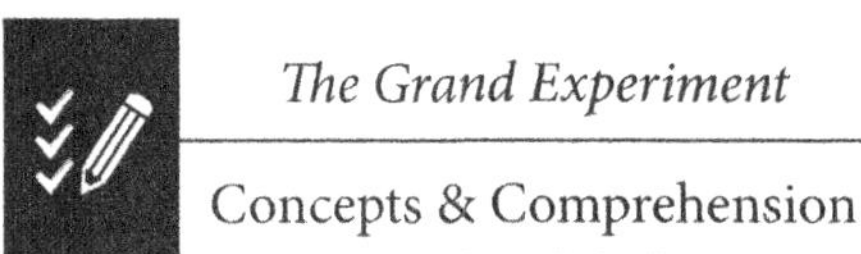

The Grand Experiment / Concepts & Comprehension	Chapter 7 Test	The Fossil Record of Invertebrates	Total score: ____of 100	Name

1. *True or False*: An invertebrate is an animal with a backbone.

2-8. Name seven invertebrate animals.

a.

b.

c.

d.

e.

f.

g.

9. *True or False*: The Cambrian Explosion was an underground explosion, the result of methane gas built up in the rock layers.

10. *True or False*: The fossil record of invertebrates matches Darwin's prediction.

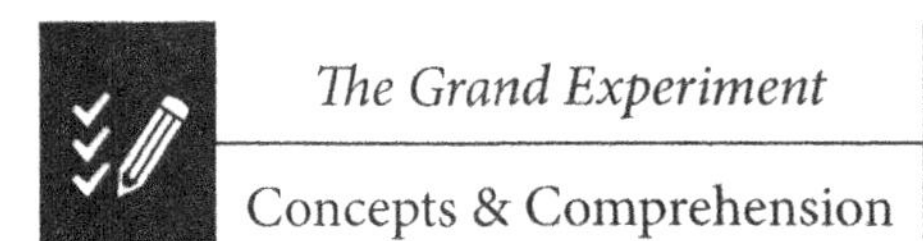

The Grand Experiment / Concepts & Comprehension	Chapter 8 Test	The Fossil Record of Fish	Total score: ____of 100	Name

1. *True or False*: Vertebrates are animals without a backbone.

2. *True or False*: Fossilized fish are rather scarce.

3. *True or False*: Fossilized fish are poorly preserved.

4. *True or False*: The theoretical evolutionary common ancestor of all fish has been discovered.

5. *True or False*: The theoretical evolutionary intermediate stages between the common ancestor of all fish and the different types of fish have been found.

6. *True or False*: According to scientists who oppose evolution, every major kind of fish appears fully formed without a trace of an ancestor.

7. *True or False*: The fossil record shows clear fossil evidence of one fish family slowly evolving into another fish family.

8. How many fossil invertebrates have been *collected* so far?

9. How many fossil fish have been *collected* by museums?

10. How many intermediate animals have been found between invertebrates and fish?

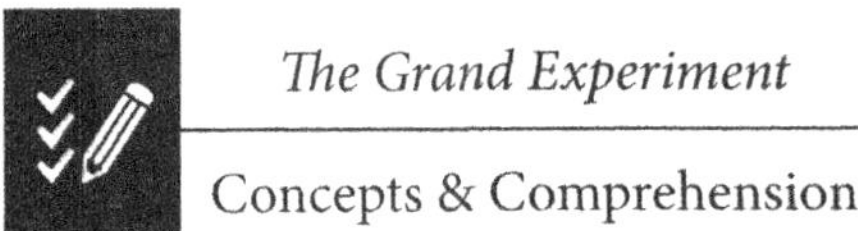

The Grand Experiment / Concepts & Comprehension	Chapter 9 Test	The Fossil Record of Bats	Total score: ____of 100	Name

1. *True or False*: All fossilized bats found so far appear fully developed and capable of flying.

2. *True or False*: The ground animal (or mammal) that bats evolved from has been discovered.

3. How many fossil bats have been found to date?

4. Have any of the theoretical evolutionary ancestors of bats been found?

5. If evolution is true and if the fossil record is complete, what should be found regarding the fossil record of bats?

6. If evolution is not true and if the fossil record is complete, what should be found regarding the fossil record of bats?

7. *True or False*: Evolution scientists have depended on speculation to decide when bats may have evolved and what happened in the evolutionary process of bat evolution.

8. *True or False*: The oldest fossil bats look similar to modern bats.

9. *True or False*: Bats first appear in the Eocene fossil layer.

10. *True or False*: Scientists who oppose the theory of evolution believe the absence of bat transitional evolutionary forms proves that evolution is not true.

1. *True or False*: Thousands of fossil sea lions have been found.

2. *True or False*: The proposed land ancestor of seals has been found.

3. Is a visible external ear present in a seal or sea lion?

4. If a land mammal (Animal A) theoretically evolved into a sea lion (Animal H) over millions of years, the theory of evolution predicts scientists would find evidence for this evolution (Animals A, B, C, D, E, F, and G) as more and more fossils are found. In the case of sea lions, which of the animals from this list of ancestors (A–G) are missing?

   (Note: The answer should be "Animals A, B, and C" or "Animals D–G," etc.)

5. Does the fossil pattern, as described in your answer for question #4, match Darwin's predictions?

6. *True or False*: Because of some similarities among seals, skunks, and otters, evolution scientists believe seals evolved from either a skunk-like or an otter-like animal.

7. *True or False*: The evolutionary ancestors of seals have been found.

8. *True or False*: Scientists who support the theory of evolution propose that sea lions evolved from a bear-like animal.

9. Which can dive deeper in the water, a seal or a nuclear submarine?

10. How long can seals hold their breath?

The Grand Experiment	Chapter 11 Test	The Fossil Record of Flying Reptiles	Total score: ____of 100	Name
Concepts & Comprehension				

1. ______________________________ are flying reptiles.

2. What are the two types of pterosaurs?

   a.

   b.

3. *True or False*: The evolutionary ancestors of pterosaurs have been found.

4. *True or False*: The fossil record of pterosaurs has a large ancestral gap.

5. *True or False*: Fossil pterosaurs have been found on only a few continents.

6. *True or False*: Pterosaurs lived during the time of the dinosaurs.

7. *True or False*: Some pterosaurs were larger than fighter jets.

8. How many fossil specimens of the (proposed) land reptile that theoretically evolved into pterosaurs have been found?

9. How many of the intermediate animals (between the land reptile that evolved into pterosaurs and actual pterosaurs) have been found?

10. How many fossil pterosaurs have been found so far?

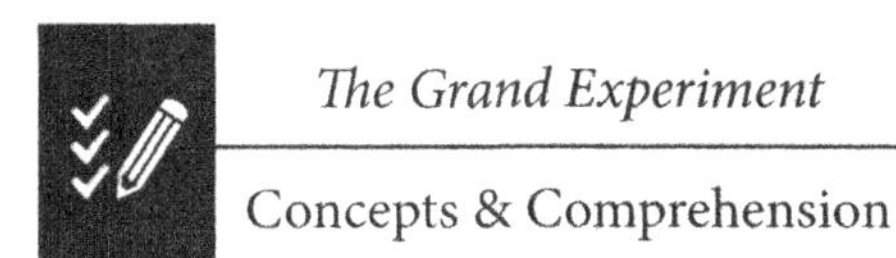

The Grand Experiment / Concepts & Comprehension	Chapter 12 Test	The Fossil Record of Dinosaurs	Total score: ____of 100	Name

1. How many *Triceratops* have been found?

2. What was the largest meat-eating dinosaur to have ever lived?

3. *True or False*: The fossil record of dinosaurs matches the predictions of the theory of evolution.

4. How many individual dinosaurs have been *collected* by museums?

5. *True or False*: Fifteen evolutionary dinosaur ancestors of *T. rex* have been discovered so far.

6. How many *T. Rex* dinosaurs have been found?

7. What is the largest and best known of the horned dinosaurs?

8. How many species of dinosaurs are known?

9. Has the common ancestor to all dinosaurs been found?

10. How many ancestors of *Triceratops* have been found?

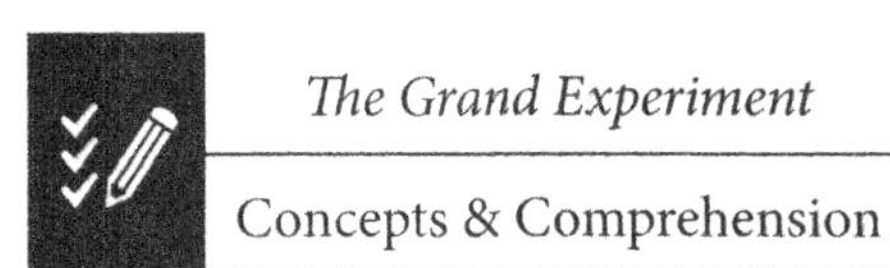

The Grand Experiment / Concepts & Comprehension	Chapter 13 Test	The Fossil Record of Whales	Total score: ____of 100	Name

1. *True or False*: Cetaceans are a group of land mammals.

2. *True or False*: Different evolution scientists express different ideas concerning which land mammal became a whale.

3. *True or False*: At the University of Michigan, one problem in the sequence of fossils demonstrating the evolution of whales is the tail of the animal called *Rodhocetus*.

4. *True or False*: A scientist can tell if an animal had a fluke by looking at the bones of the tail.

5. *True or False*: Some *evolution* scientists now suggest *Ambulocetus* may not be a direct ancestor to whales because of the strange location of its eyes.

6. According to Dr. Gingerich, what body feature is similar between hoofed hyenas and archaic whales?

7–9. Name three possible ancestors of whales as proposed by modern evolution scientists.

a.

b.

c.

10. Does Dr. Gingerich, who discovered *Rodhocetus,* still believe it had a whale's tail and flippers?

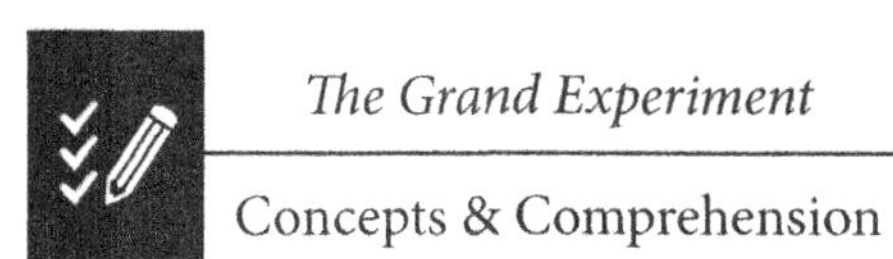

The Grand Experiment / Concepts & Comprehension	Chapter 14 Test	The Fossil Record of Birds —Part 1: *Archaeopteryx*	Total score: ____of 100	Name

1. *True or False*: Superficially, *Archaeopteryx* looks like a modern bird.

2. Fossil *Archaeopteryx* birds have only been found in what country?

3. How many specimens of *Archaeopteryx* have been found over the past 140 years?

4. Scientists and museums often portray which dinosaur species as being the closest dinosaur that may have evolved into birds?

5. *True or False*: Modern-appearing animals have been found with *Archaeopteryx.*

6. *True or False*: Evolution scientists cannot agree on what type of reptile evolved into birds.

7. *True or False*: According to Dr. Peter Wellnhofer, an expert who has worked with three of the original *Archaeopteryx* fossils, this animal (*Archaeopteryx*) had a scaly head, similar to a dinosaur or snake.

8. How many fossil birds have been *collected* by museums?

9. Fossils from how many dinosaurs have been *collected* by museums?

10. *True or False*: Birds are the only types of vertebrates (animals with backbones) with claws on their wings.

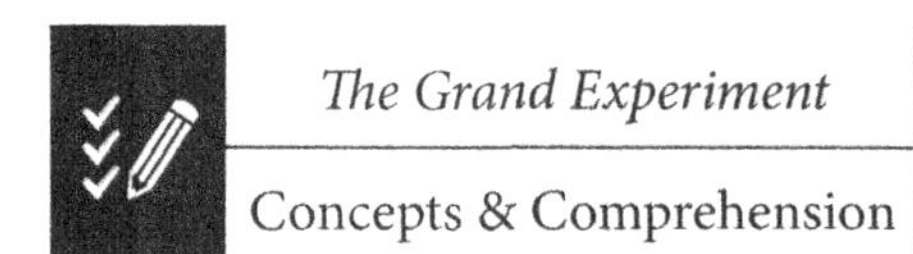

The Grand Experiment / Concepts & Comprehension	Chapter 15 Test	The Fossil Record of Birds —Part 2: Feathered Dinosaurs	Total score: ____of 100	Name

1. *True or False*: Some evolution scientists believe the "feathered dinosaurs" from China are not dinosaurs at all; rather, they believe they are simply flightless birds.

2. In what decade were the Chinese fossils found?

3. Of the 88 bones comprising the fossil "feathered dinosaur" *Archaeoraptor liaoningensis*, how many were substituted from other animals?

4. How many different animals contributed bones to the "feathered dinosaur" *Archaeoraptor liaoningensis*?

5. *True or False*: According to Dr. Rowe, the perpetrator of the fake fossil, *Archaeoraptor liaoningensis*, may have been a scientist.

6. *True or False*: A prominent scientist has accused *National Geographic* of placing feathers on *Tyrannosaurus rex*, even though no feathers had been found with the fossils of this dinosaur.

7. *True or False*: *National Geographic* magazine published a seven-page article explaining the mistakes made in the original *Archaeoraptor liaoningensis* story they had published.

8. When was the CT scan done on *Archaeoraptor liaoningensis* showing it was a fake? (month/year)

9. When did *National Geographic* publish its story proclaiming that *Archaeoraptor liaoningensis* was a "flying dinosaur"? (month/year)

10. *True or False*: Dr. Wellnhofer from the University of Texas did a CT scan on *Archaeoraptor liaoningensis* and found it was a fake.

The Grand Experiment / Concepts & Comprehension	Chapter 16 Test	The Fossil Record of Flowering Plants	Total score: ____of 100	Name

1. *True or False*: Some scientists believe the fossil record is more than adequate to demonstrate that the evolution of plants did *not* occur.

2. *True or False*: Evolution scientists speak in terms of "mystery" when talking about the origin of flowering plants.

3. How many species of *flowering* plants are alive today?

4. How many fossil plants are in museums today?

5–10. List six flowering plant structures (such as a branch) that have been found as fossils. List from the smallest in size to the largest in size.

a.

b.

c.

d.

e.

f.

The Grand Experiment / Concepts & Comprehension	Chapter 17 Test	The Origin of Life — Part 1: The Formation of DNA	Total score: ____of 100	Name

1. According to evolution scientists, how long ago did life begin?

2. *True or False*: Organic chemicals have successfully been formed in the laboratory.

3. All living organisms today are comprised of these three necessary components:
   a.

   b.

   c.

4. _____________ contains the genetic blueprint of life.

5. How many letters of DNA are required to instruct the cell to place a single amino acid into a protein chain?

6. Do long strands of spiraled DNA form naturally?

7. A ________ -shaped structure is important for DNA because it compacts and protects the DNA.

8. The theoretical origin of the first living single-cell organism or bacteria-like organism is what scientists mean when they refer to the __________ of __________.

9. Name the four DNA letters. _____ _____ _____ _____

10. *True or False*: If DNA cannot form spontaneously from chemicals in the proper length, order, and shape, then life cannot begin.

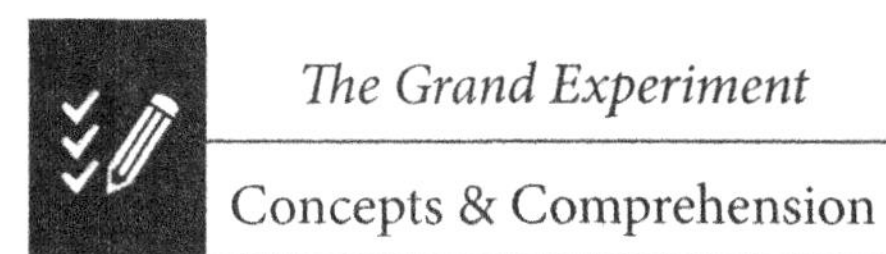

The Grand Experiment / Concepts & Comprehension	Chapter 18 Test	The Origin of Life — Part 2: The … of Proteins	Total score: ____of 100	Name

1. *True or False*: Conceptually, DNA looks like a chain.

2. __________ __________ make up the individual links of a protein chain.

3. Do proteins form naturally in water?

4. How many unique proteins are in the simplest bacterium *living today*?

5. What is the substance called that is created by the unnatural congealing of amino acids using heat?

6. Do scientists who oppose evolution agree with the idea of a proteinoid?

7. How many different *kinds* (types) of amino acids are found in living organisms?

8. Why is it a problem to believe that the first proteins began in the ocean?

9. Write out the formula for the number of amino acids needed to form the simplest first single-cell bacterium-like organism.

10. *True or False*: Proteinoids have been observed to form in nature.

The Grand Experiment / Concepts & Comprehension	Chapter 19 Test	The Origin of Life — Part 3: The … of Amino Acids	Total score: ____of 100	Name

1. What is the name of the scientist who successfully formed some amino acids in his laboratory under "natural settings"?

2. What year did he do this?

3. Name any two of the five components of his apparatus.

   a.

   b.

4. What is the main criticism of his experiment?

5. Did his apparatus produce left or right-handed amino acids or both?

6. What forms of amino acids are poisonous to all living systems?

7. *True or False*: Oxygen in the atmosphere, in the form of ozone, is necessary to protect life but oxygen on earth prevents amino acids forming from a mixture of chemicals.

8. All living organisms use only what form of amino acids?

9. Have scientists ever produced a single-cell organism from a mixture of chemicals in a laboratory?

10. What occurrence in nature did Dr. Miller suggest the tungsten electrode in his device represented?

Chapter Tests

for Use with

Living Fossils

Living Fossils / Concepts & Comprehension	Chapter 1 Test	The Challenge That Would Change My Life	Total score: ____of 100	Name

Outline Darwin's theory of evolution using only 10 words.

1.

2.

3.

4.

5.

6.

7.

8.

9.

10.

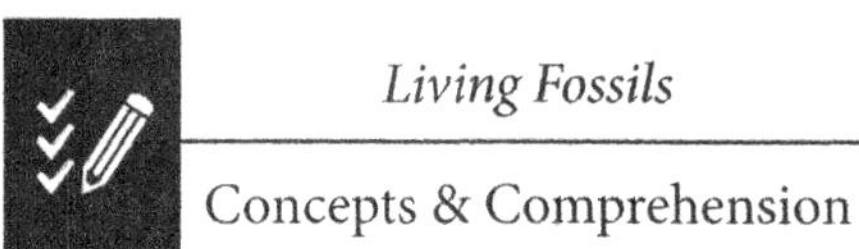

Living Fossils / Concepts & Comprehension	Chapter 2 Test	How Can You Verify Evolution?	Total score: ____of 100	Name

1–4. Name the four most well-recognized examples of living fossils from the dinosaur layers.

a.

b.

c.

d.

5. What question did the author want to ask to test evolution as he traveled from dig site to dig site around the world?

6-7. What two predictions did the author make — ideas opposite of the theory of evolution — in order to test the theory of evolution?

a.

b.

8-9. Name two scientific ideas that were believed by scientists for years but were later shown to be false.

a.

b.

10. *True or False*: A scientist tests a theory by vigorously trying to prove it true.

Living Fossils / Concepts & Comprehension	Chapter 3 Test	The Naming Game	Total score: ____of 100	Name

1-10. Fill in the blanks (such as genus, class, order, phylum, etc.) for classifying a dog:

Kingdom	Animal
Species	Familiaris

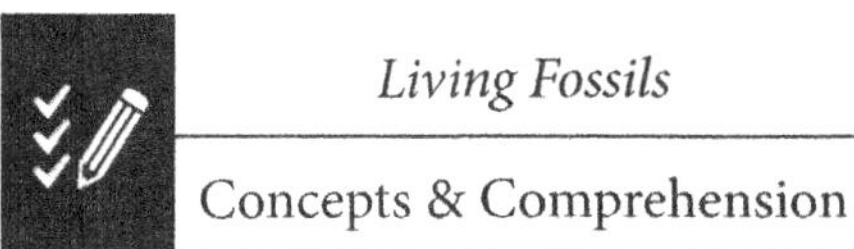

Living Fossils / Concepts & Comprehension	Chapter 4 Test	Echinoderns	Total score: ____of 100	Name

1. In which animal phylum do starfish belong?

2. *True or False*: Echinoderms have 6-fold symmetry.

3. *True or False*: Sea urchins look like a banana.

4-5. List the two types of crinoids.

a.

b.

6-10. Name five classes of echinoderms that have been found in dinosaur rock layers that appear similar to modern forms.

a.

b.

c.

d.

e.

Living Fossils / Concepts & Comprehension	Chapter 5 Test	Aquatic Arthropods	Total score: ____of 100	Name

1-5. Name five major types of aquatic arthropods living today that have been found in dinosaur layers.

a.

b.

c.

d.

e.

6-8. What three body features do arthropods have?

a.

b.

c.

9–10. Name the bird and dinosaur found in Solnhofen, Germany, along with modern-appearing aquatic arthropods.

a.

b.

Living Fossils	Chapter 6 Test	Land Arthropods	Total score: ____of 100	Name
Concepts & Comprehension				

1. *True or False*: All of the major insect orders living today have been found in dinosaur rock layers.

2. *True or False*: Both of the major Myriapod orders living today (centipedes and millipedes) have been found in dinosaur rock layers.

3. What percentage of the living land Arachnid orders have also been found in dinosaur rock layers?

4-9. List six examples of land arthropods.

a.

b.

c.

d.

e.

f.

10. *True or False*: The author's experiment predicted that he would find modern-appearing land arthropods in dinosaur rock layers if evolution was *not* true.

Living Fossils	Chapter 7 Test	Bivalve Shellfish	Total score: ____of 100	Name
Concepts & Comprehension				

1. *True or False*: Bivalve shellfish have two shells.

2-5. Name four common bivalve shellfish living today.

a.

b.

c.

d.

6. What bivalve shellfish is the icon for the Shell Oil Company?

7-10. Name four common bivalve shellfish living today that were also found in dinosaur rock layers.

a.

b.

c.

d.

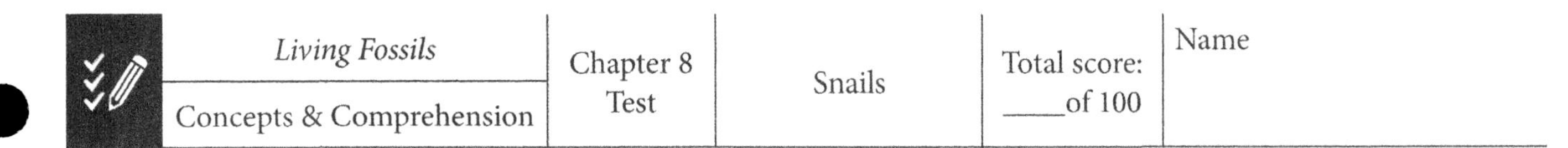

Living Fossils
Concepts & Comprehension
Chapter 8
Test
Snails
Total score:
____of 100
Name

True or False:

1. Snails have one shell whereas bivalves have three.

2. Snails and bivalves belong to the same phylum.

3. Modern-appearing freshwater snails were found at Dinosaur National Monument.

4. A modern-appearing saltwater snail was found at the same location as the dinosaur *Coelophysis.*

5. Modern-appearing slit shells and moon snails were found in an Australian dinosaur rock layer.

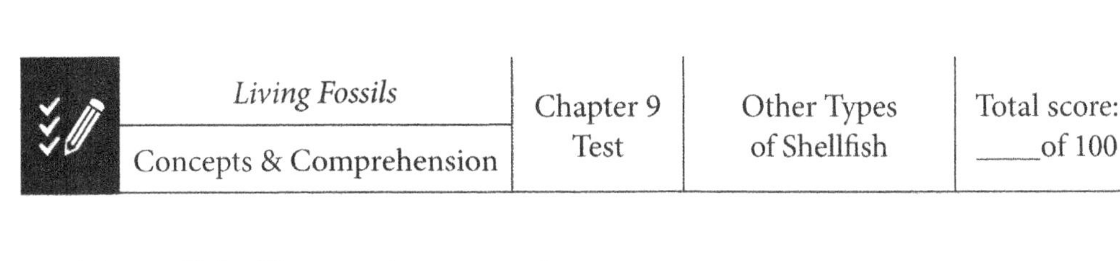

Living Fossils
Concepts & Comprehension
Chapter 9
Test
Other Types
of Shellfish
Total score:
____of 100
Name

1. *True or False*: The nautilus is a snail.

2. *True or False*: Tusk shells look like elephant tusks and are just as large.

3. *True or False*: Modern-appearing shellfish from all five major classes of shellfish living today in the Phylum Mollusca have been found in dinosaur rock layers.

4. How many plates do sea cradles have?

5. *True or False*: The siphuncle of the living nautilus acts as a mouth and is used to eat food.

Match these five classes from the Phylum Mollusca to their common names.

6. Bivalves — a. Class Polyplacophora

7. Snails — b. Class Scaphoda

8. Chambered Shellfish — c. Class Gastropoda

9. Tusk Shells — d. Class Cephalopoda

10. Sea Cradles — e. Class Bivalvia

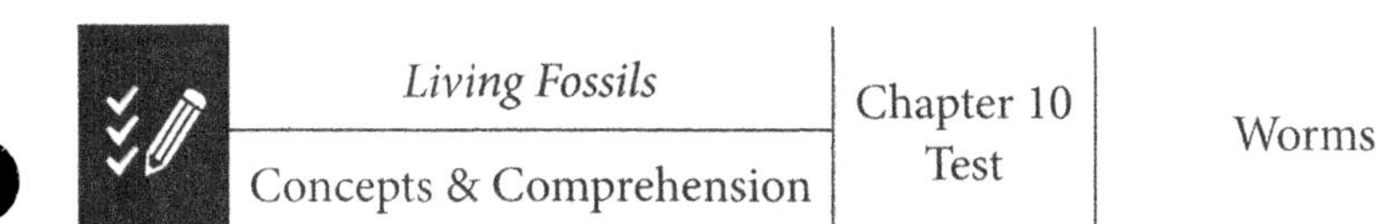

Chapter 10 Test	Worms	Total score: ____of 100	Name

True or False:

1. Earthworms belong to Phylum Arthropoda.

2. Tube worms live in the ground.

3. There are 14 worm phyla groups, but most are obscure.

4. Most people have seen examples of only one living phyla of worm, the segmented earthworms, Phylum Annelida.

5. Examples from both classes of segmented worms living today have also been found as fossils in dinosaur rock layers.

Living Fossils	Chapter 11 Test	Sponges and Corals	Total score: ____of 100	Name
Concepts & Comprehension				

1. *True or False*: Sponges are members of Phylum Cnidaria.

2. *True or False*: Modern-appearing sponges and corals have been found as fossils in dinosaur rock layers.

3. *True or False*: Of the two types of corals living today, hard corals and soft corals, only soft corals have been found in dinosaur rock layers.

4. *True or False*: All three classes of sponges living today were also alive during the time of the dinosaurs.

5. *True or False*: There are over 30 animal phyla living today.

6-8. List the three classes of sponges living today. May use formal name or informal name of classes.

a.

b.

c.

9-10. List the two types of corals living today.

a.

b.

Living Fossils / Concepts & Comprehension	Chapter 12 Test	Bony Fish	Total score: ____of 100	Name

1-7. List seven modern types of bony fish found with dinosaurs.

a.

b.

c.

d.

e.

f.

g.

8. What did the author predict he would find in dinosaur rock layers regarding bony fish if evolution was *not* true?

9. Which class of fish living today is the largest group?

10. What phylum do fish belong to?

Living Fossils	Chapter 13 Test	Cartilaginous Fish	Total score: ____of 100	Name
Concepts & Comprehension				

1-3. Three cartilaginous fish were found in Solnhofen, Germany, along with a dinosaur and the famous toothed bird *Archaeopteryx.* What types of fish did these fossil fish look like?

a.

b.

c.

4. Shark teeth were found in New Mexico dinosaur rock layers. These shark teeth look similar to the teeth of what modern shark?

5. *True or False*: The author predicted he would find *modern* types of fish in dinosaur layers if evolution was *not* true and this was what he found.

Living Fossils / Concepts & Comprehension	Chapter 14 Test	Jawless Fish	Total score: ____of 100	Name

1-2. List the two types of jawless fish living today.

a.

b.

3. *True or False*: The two types of jawless fish living today also lived during the time of the dinosaurs and appear almost unchanged since the dinosaur era.

4. *True or False*: The author predicted that if evolution is *not* true, then modern fish should be found in dinosaur rock layers.

5. *True or False*: There are three classes of fish living today and all three have been found in dinosaur rock layers and appear basically unchanged from modern forms.

Living Fossils	Chapter 15 Test	Amphibians	Total score: ____of 100	Name
Concepts & Comprehension				

True or False:

1. A fossil toad was found in a coal mine in Bernissart, Belgium, along with 29 dinosaurs.

2. The dinosaur-era salamander called *Karaurus* was built like a modern salamander.

3. Fossil frogs have been found in Jurassic rock layers.

4. Frogs are members of the Phylum Arthropoda.

5. Both *major* groups of amphibians living today were also alive during the time of the dinosaurs.

Living Fossils	Chapter 16 Test	Crocodilians	Total score: ____of 100	Name
Concepts & Comprehension				

1. *True or False*: Crocodiles and alligators are members of the Phylum Chordata.

2. *True or False:* Fossilized alligator and crocodile skulls were found at Dinosaur Provincial Park in Alberta, Canada.

3-5. Name three types of crocodilians that are alive today that were also alive during the time of the dinosaurs.

a.

b.

c.

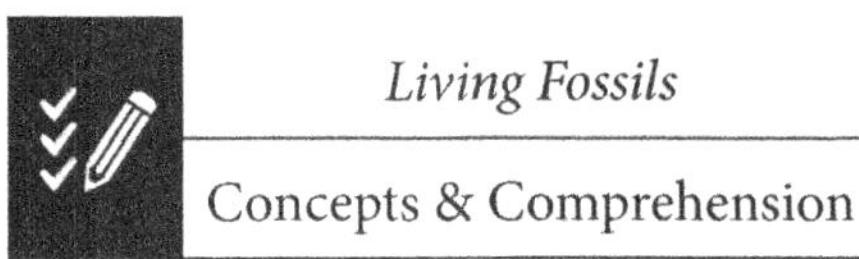

	Chapter 17 Test	Snakes	Total score: ____of 100	Name
Living Fossils / Concepts & Comprehension				

1. *True or False*: Snakes belong to the Phylum Chordata.

2. *True or False*: Snakes have never been found in dinosaur rock layers.

3. *True or False*: A modern-appearing boa constrictor was found at a dinosaur dig site in Montana along with a *Triceratops*.

4-10. Modern-appearing animals from all of the seven major animal phyla have been found in dinosaur rock layers. What are these seven phyla?

a.

b.

c.

d.

e.

f.

g.

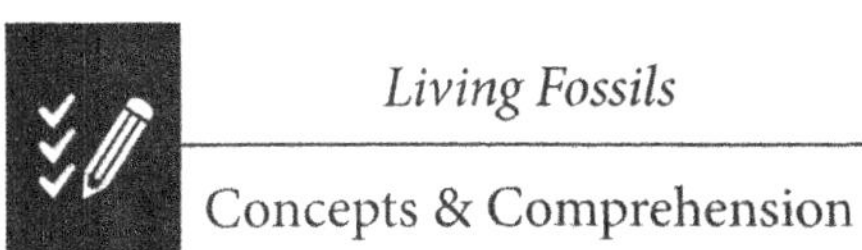

Living Fossils / Concepts & Comprehension	Chapter 18 Test	Lizards	Total score: ____of 100	Name

1-4. List four types of lizards living today that have been found in dinosaur rock layers.

a.

b.

c.

d.

5. What phylum includes lizards?

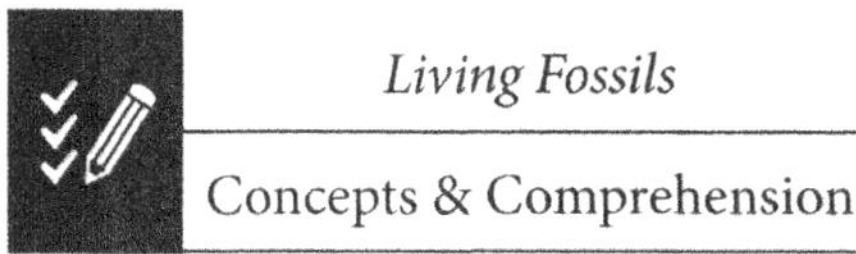

Living Fossils / Concepts & Comprehension	Chapter 19 Test	Turtles	Total score: ____of 100	Name

True or False:

1. Both sub-orders of turtles living today have been found in dinosaur rock layers.

2. Box turtles have been found in dinosaur rock layers.

3. All four groups of reptiles living today have been found in dinosaur rock layers.

4. Fossils of all five groups of vertebrates living today have been found in dinosaur rock layers and appear basically unchanged.

5. Fossils of all seven animal phyla groups living today have been found in dinosaur rock layers and appear basically unchanged.

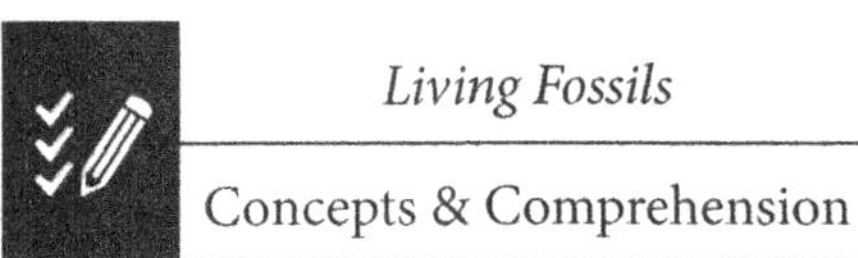

Living Fossils / Concepts & Comprehension	Chapter 20 Test	Birds	Total score: ____of 100	Name

1-10. List ten modern types of birds that have been found in dinosaur rock layers.

a.

b.

c.

d.

e.

f.

g.

h.

i.

j.

Living Fossils / Concepts & Comprehension	Chapter 21 Test	Mammals	Total score: ____of 100	Name

True or False:

1. All three subclasses of mammals living today have been found in dinosaur rock layers.

2. Modern-appearing mammals have been found in dinosaur rock layers.

3. Mammals were found in dinosaur rock layers long before Darwin wrote *The Origin of Species.*

4. More than 300 genera of mammals have been found in dinosaur rock layers and even more *species* of mammals have been found in dinosaur rock layers.

5. Only 3 *complete* mammal skeletons have been found in dinosaur rock layers.

6. Evolution scientists, such as Dr. Luo of the Carnegie Museum, suggest that the term "Age of the Reptiles," when referring to the time of the dinosaurs, is a misnomer.

7. For over a hundred years, scientists who supported evolution suggested that the largest mammal living during the time of the dinosaurs was about the size of a mouse or rat.

8. Mammals the size of a 30-pound collie have been found with dinosaurs.

9. Reconstructions of the mammal *Gobiconodon* look like a modern hedgehog.

10. A mammal which looks like a ringtail opossum has been found in dinosaur rock layers.

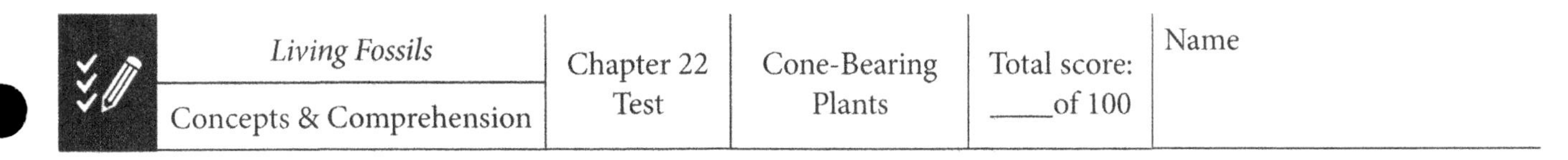

Living Fossils / Concepts & Comprehension	Chapter 22 Test	Cone-Bearing Plants	Total score: ____of 100	Name

1. What is the plant division that contains pine trees and sequoias?

2. *True or False*: Modern-appearing sequoias have been found in dinosaur rock layers.

3. *True or False*: There are seven major animal phyla living today, but only four of these have been found with dinosaurs.

4. *True or False*: All three divisions of cone-bearing plants, grouped together, are called gymnosperms.

5-10. Name six types of cone-bearing plants that are alive today that have been found in dinosaur rock layers from Division Coniferae, Division Cycadophyta, or Division Ginkgophyta.

a.

b.

c.

d.

e.

f.

Living Fossils	Chapter 23 Test	Spore-Forming Plants	Total score: ____of 100	Name
Concepts & Comprehension				

1-3. Match the common and official names for these groups of spore-forming plants.

1. Moss
2. Club moss
3. Ferns and horsetails

A. Division Pteridophyta
B. Division Lycopodiophyta
C. Division Bryophyta

4. *True or False*: Only two of the three divisions of spore-forming plant divisions living today have also been found in dinosaur rock layers.

5. What is the genus name for peat moss?

Living Fossils / Concepts & Comprehension	Chapter 24 Test	Flowering Plants	Total score: ____of 100	Name

1-10. List ten common flowering plants that have been found in dinosaur rock layers.

a.

b.

c.

d.

e.

f.

g.

h.

i.

j.

Living Fossils	Chapter 25 Test	Coming Full Circle — My Conclusions	Total score: ____of 100	Name
Concepts & Comprehension				

True or False:

1. The author had access to 5% of the world's fossil collections to carry out this experiment.

2. When the author ignored the subjective genus and species names he concluded that evolution is *not* true.

3. Despite visiting over 60 museums, the author only saw two exhibited specimens of the known 100 complete mammal skeletons found in dinosaur rock layers.

4. Some animals, such as dinosaurs and pterosaurs, have gone extinct, while other animals which lived alongside dinosaurs, such as mammals, birds, lizards, and shellfish, have not gone extinct.

5. All seven *major* plant divisions living today and all seven *major* animal phyla living today have been found in dinosaur rock layers, and many of these dinosaur-era organisms appear nearly the same as those living today.

Sectional & Final Exams

for Use with

Evolution: The Grand Experiment

	*The Grand Experiment* / Concepts & Comprehension	Sectional Exam 1	Scope: Chapters 1–3	Total score: ____of 100	Name

Answer the following questions. Each answer is worth 3 points.

1. What are the two opposing views on the origin of life?

   a.

   b.

2. Name one of the four best evidences *against* evolution, cited by scientists who oppose the theory.

3. What famous artist painted a scene of a deity creating man on the ceiling of the Sistine Chapel?

4. Since Darwin first published his theory of evolution, how many fossils have been collected by museums?

5. Since the middle of the 20th century, there have been a growing number of scientists who reject the theory of evolution based on the discovery of ____________ and ____________, of which Darwin was unaware.

6. What theory eventually replaced the theory of spontaneous generation and is the basis for the modern theory of how life began naturally?

7. Pond scum that forms in a bottle of boiled pond water is actually a collection of ___________.

8. *True or False*: The theory of spontaneous generation was believed by scientists for less than 1,000 years.

9. *True or False*: Cheesecloth prevents flies from laying their eggs on meat.

10. *True or False*: Dr. Louis Pasteur disproved the spontaneous generation of bacteria.

11. Before Darwin's theory of evolution, what theory did scientists believe explained the origin of life?

12. In what year did Darwin publish his first book on evolution?

13. Define the theory of acquired characteristics.

14. *True or False*: The modern theory of evolution proposes that all forms of life ultimately evolved from a single-celled organism or bacterium-like organism.

15. *True or False*: We now know that changes in the cells of the body (e.g., changes in the skin cells and muscle cells) that occur during life can be passed on to the next generation.

16. List, in order, the eight types of animals or organisms that theoretically evolved from one another, from the theoretical primordial form to a human being, according to the modern theory of evolution.

    a.

    b.

    c.

    d.

    e.

    f.

    g.

    h.

17. According to the theory of evolution, how long did it take to accomplish the evolution from a single-cell organism to a human (general answer)?

18. What was the name of the 17th-century scientist who "proved" that mice spontaneously generate from wheat and dirty underwear?

19. How is spontaneous generation an example of an idea that leads to bad science?

20. Use the example of a horse and a giraffe to explain the theory of acquired characteristics that some scientists felt would mean we could see a horse with a very long neck. What was the flaw in their reasoning?

Short Answer - Applied Knowledge (13 Points)

21. Which of the two explanations for how life began is one that you had before this course began? How has what you have learned so far changed your perception of this controversial issue?

The Grand Experiment	Sectional Exam 2	Scope: Chapters 4-9	Total score: ____of 100	Name
Concepts & Comprehension				

Each question is worth 3 points. Final short essay question is worth 10 points.

1. What is possibly the largest animal to have ever lived on earth?

2. Charles Darwin reasoned that whales could have evolved from what land mammal?

3. Today scientists recognize that an animal could only evolve by accidental ______________ in the DNA of reproductive cells.

4. What new trait did Dr. Morgan observe that made him realize that accidental mutations accounted for new traits?

5. What four letters make up DNA?

   a.

   b.

   c.

   d.

6. *True or False*: The animal order called Cetacea is made up of fish, whales, and dolphins.

7. *True or False*: Using just artificial breeding or just natural selection, completely new body parts can be added to a breed of plants or animals.

8. Write out the definition of "convergent evolution."

9. Why do scientists who oppose evolution object to using similarities as "proof" of evolution?

10. Name three animals that have wings but are unrelated.

    a.

    b.

    c.

11. Name two similarities between the giant panda and the red panda that prompted scientists to (incorrectly) assume they were related to each other.

    a.

    b.

12. Describe how a hyrax is similar to a horse and a rhinoceros.

13. What two animals do evolution scientists believe seals may have evolved from?

    a.

    b.

14. Name two animals with finned feet and front flippers that are not closely related to each other.

    a.

    b.

15. *True or False*: Darwin recognized that the fossil record did not match what his theory predicted.

16. *True or False*: Since Darwin's time, millions of (difficult-to-fossilize) soft-bodied plants and animals have been found as fossils.

17. *True or False*: If evolution is true and if the fossil record is nearly complete, the fossil record should show one animal slowly changing into a completely different type of animal over time.

18. What would the fossil record show if the fossil record was nearly complete and if evolution was not true?

19. Give five examples of soft-bodied animals or plants that have been found as fossils.

    a.

    b.

    c.

    d.

    e.

20. *True or False*: An invertebrate is an animal with a backbone.

21. Name seven invertebrate animals.

    a.

    b.

    c.

    d.

    e.

    f.

    g.

22. Name the oldest three fossil layers in order (according to the theory of evolution).

    a.

    b.

    c.

23. *True or False*: Scientists who *support* the theory of evolution believe that the absence of predicted ancestors for trilobites, jellyfish, and sea pens is because the ancestors were soft-bodied and soft-bodied animals were not as likely to have been fossilized.

24. *True or False*: Scientists who *oppose* the theory of evolution believe that the absence of predicted ancestors for trilobites, jellyfish, and sea pens is because the ancestors were soft-bodied and soft-bodied animals were not as likely to have been fossilized.

25. List three *fossilized* soft-bodied invertebrates that have been found.

    a.

    b.

    c.

26. *True or False*: The theoretical evolutionary common ancestor of all fish has been discovered.

27. *True or False*: According to scientists who oppose evolution, every major kind of fish appears fully formed without a trace of an ancestor.

28. *True or False*: According to Dr. Long, proponent of and expert in fish evolution, the evidence for how sharks evolved is overwhelming.

29. *True or False*: All fossilized bats found so far appear fully developed and capable of flying.

30. *True or False*: The ground animal (or mammal) that bats evolved from has been discovered.

31. Why are intermediate animals so important to both creation scientists and secular scientists? How does the number of confirmed transitional fossils from among the hundreds of millions of documented fossils found so far – if any – help to answer this question?

	The Grand Experiment	Sectional Exam 3	Scope: Chapters 10–15	Total score: ____of 100	Name
	Concepts & Comprehension				

Each question is worth 2 points. Final short essay question is worth 10 points.

1. *True or False*: Thousands of fossil sea lions have been found.

2. If a land mammal (Animal A) theoretically evolved into a sea lion (Animal H) over millions of years, the theory of evolution predicts scientists would find evidence for this evolution (Animals A, B, C, D, E, F, and G) as more and more fossils are found. In the case of sea lions, which of the animals from this list of ancestors (A–G) are missing? ____________________ (Note: The answer should be "Animals A, B, and C" or "Animals D–G," etc.)

3. What distinguishes sea lions from seals?

4. The California sea lion, the Australian sea lion, and the Stellar sea lion are all various forms of ___________ __________. (Be as specific as possible. The answer is not aquatic mammals or pinnipeds.)

5. *True or False*: Scientists who support the theory of evolution propose that sea lions may have evolved from a bear-like animal.

6. What are the two types of pterosaurs?

   a.

   b.

7. *True or False*: The fossil record of pterosaurs has a large ancestral gap.

8. How many fossil specimens of the (proposed) land reptile that theoretically evolved into pterosaurs have been found?

9. How many of the intermediate animals (between the land reptile that evolved into pterosaurs and actual pterosaurs) have been found?

10. *True or False*: The fossil record of pterosaurs does not match the predictions of the theory of evolution.

11. *True or False*: The fossil record of pterosaurs is very poor, with very little known about their wings.

12. *True or False*: Because pterosaurs are extinct, this proves evolution and the idea that animals changed over time.

13. How many *Triceratops* have been found?

14. *True or False*: Fifteen evolutionary dinosaur ancestors of *T. rex* have been discovered so far.

15. How many species of dinosaurs are known?

16. Has the common ancestor to all dinosaurs been found?

17. What are the names of the two types of dinosaurs (based on the shape of the pelvis)?

    a.

    b.

18. What is the largest of the horned dinosaurs?

19. *True or False*: *Triceratops* is a member of the Ornithischian group of dinosaurs.

20. *True or False*: *T. rex* was the largest meat-eating dinosaur to have ever lived.

21. *True or False*: Cetaceans are a group of land mammals.

22. *True or False*: Different evolution scientists express different ideas concerning which land mammal became a whale.

23. *True or False*: A scientist can tell if an animal had a fluke by looking at the bones of the tail.

24. *True or False*: *Ambulocetus* was considered a whale only because evolution scientists believed it evolved into a whale, not because it had a blowhole or a whale's tail.

25. *True or False*: Some evolution scientists have put *Basilosaurus* on the evolutionary line to modern whales even though other evolution scientists believe this is not true.

26. *True or False*: According to scientists who support evolution, whale evolution is considered to be one of the best fossil proofs for evolution.

27. *True or False*: Scientists who support the idea that whales evolved from a land mammal cannot come to a consensus regarding which land mammal it was.

28. What land animal did Charles Darwin believe could have evolved into a whale?

29. There are two major problems with hippos being the evolutionary ancestor of whales. What are these two problems?

    a.

    b.

30. *True or False*: Superficially, *Archaeopteryx* looks like a modern bird.

31. Fossil *Archaeopteryx* birds have only been found in what country?

32. How many specimens of *Archaeopteryx* have been found over the past 140 years?

33. Scientists and museums often portray which dinosaur species as being the closest dinosaur that may have evolved into birds?

34. *True or False*: Modern-appearing animals have been found with *Archaeopteryx.*

35. *True or False*: The first fossil specimen of *Archaeopteryx* was found in the nineteenth century.

36. *True or False*: The dinosaur *Deinonychus* lived before the bird *Archaeopteryx.*

37. *True or False*: According to scientists who oppose evolution, only a whole series of fossils, showing a dinosaur slowly changing into a flying bird, such as *Archaeopteryx*, would prove the evolution of birds.

38. *True or False*: Some evolution scientists believe the "feathered dinosaurs" from China are not dinosaurs at all; rather, they believe they are simply flightless birds.

39. In what decade were the Chinese fossils found?

40. *True or False*: *National Geographic* magazine published a seven-page article explaining the mistakes made in the original *Archaeoraptor liaoningensis* story they had published.

41. *True or False*: The feathers of the Chinese specimens are asymmetric, suggesting they could fly.

42. *True or False*: Feathers of modern birds that cannot fly are asymmetric.

43. *True or False*: The Chinese "feathered dinosaurs" are 4 to 5 feet long, the size of a small dinosaur.

44. *True or False*: Although the fossil *Confuciusornis* looks authentic and solid, it is actually a composite fossil made up of mortar, paint, and substitute rock backing.

45. *True or False*: According to an evolution scientist from the Smithsonian Museum of Natural History, *National Geographic* has made "melodramatic assertions" and participated in "hype" regarding their support of the idea that birds evolved from dinosaurs.

Short Answer — Applied Knowledge

46. Based on the information you have read in this section of the book, how would you present an argument for or against the popular conception of the evolution of whales? Be sure to include at least four short points regarding any evidences, questions, and uncertainties.

	The Grand Experiment	Sectional Exam 4	Scope: Chapters 16–19	Total score: ____of 100	Name
	Concepts & Comprehension				

Each question is worth 3 points. Final short essay question is worth 10 points.

1. *True or False*: Some scientists believe the fossil record is more than adequate to demonstrate that the evolution of plants did not occur.

2. *True or False*: Evolution scientists speak in terms of "mystery" when talking about the origin of flowering plants.

3. How many species of flowering plants are alive today?

4. How many fossil plants are in museums today?

5. List six examples of flowering plants (such as an apple tree):

   a.

   b.

   c.

   d.

   e.

   f.

6. *True or False*: Charles Darwin could see the fossil evidence for flowering plants evolving.

7. The scientific name for flowering plants is ________________________________.

8. According to evolution scientists, how long ago did life begin?

9. *True or False*: Organic chemicals have successfully been formed in the laboratory.

10. All living organisms today are comprised of these three necessary components:

a.

b.

c.

11. ______________________________ contains the genetic blueprint of life.

12. *True or False*: The proposed earliest form of life, a single-cell bacteria-like organism, has never been (re)produced in a laboratory using chemicals.

13. Which is more likely, winning the National Powerball Lottery every day for 365 days in a row or DNA forming naturally to create life?

14. How many letters of DNA form in a natural strand of DNA?

15. How many letters of DNA would be needed to form just one protein, 300 amino acids long?

16. What is the purpose of spiraling DNA?

17. *True or False*: Conceptually, proteins look like a chain.

18. ______________________________ ______________________________ make up the individual links of a protein chain.

19. Do proteins form naturally in water?

20. How many unique proteins are in the simplest bacterium living today?

21. What is the substance called that is created by the unnatural congealing of amino acids using heat?

22. Write out the calculation (number of proteins x the number of amino acids) necessary to form the simplest single-celled organism.

23. How many different kinds (types) of amino acids are found in living organisms today?

24. What is the consequence of incorrectly placing one amino acid in a protein chain?

25. *True or False*: Oxygen in the atmosphere, in the form of ozone, is necessary to protect life but oxygen on earth prevents amino acids forming from a mixture of chemicals.

26. All living organisms use only what form of amino acids?

27. *True or False*: Modern scientists have formed a single-cell organism in a laboratory by mixing chemicals together.

28. *True or False*: The Stanley Miller device produced only left-handed amino acids.

29. List three basic functions of proteins.

    a.

    b.

    c.

30. What three components are necessary for life to begin? (Name the three components of the theoretical first single-celled, bacterium-like organism.)

    a.

    b.

    c.

31. List three diseases in humans resulting from the placement of just one incorrect amino acid in a protein chain.

a.

b.

c.

	The Grand Experiment	Final Comprehensive Exam 1	Scope: Chapters 1–19	Total score: ____of 100	Name
	Concepts & Comprehension				

Each question is worth 2 points.

1. What are the two opposing views on the origin of life?

2. Name one of the four best evidences against evolution, cited by scientists who oppose the theory.

3. What famous artist painted a scene of a deity creating man on the ceiling of the Sistine Chapel?

4. Since Darwin first published his theory of evolution, how many fossils have been collected by museums?

5. Since the middle of the 20th century, there have been a growing number of scientists who reject the theory of evolution based on the discovery of _____________ and _____________, of which Darwin was unaware.

6. What theory eventually replaced the theory of spontaneous generation and is the basis for the modern theory of how life began naturally?

7. Pond scum that forms in a bottle of boiled pond water is actually a collection of ____________.

8. *True or False*: The theory of spontaneous generation was believed by scientists for less than 1,000 years.

9. *True or False*: Cheesecloth prevents flies from laying their eggs on meat.

10. *True or False*: Dr. Louis Pasteur disproved the theory of spontaneous generation of bacteria.

11. Name three animals that have wings but are unrelated.

    a.

    b.

    c.

12. Name two similarities between the giant panda and the red panda that prompted scientists to (incorrectly) assume they were related to each other.

    a.

    b.

13. Describe how a hyrax is similar to a horse and a rhinoceros.

14. What two animals do evolution scientists believe seals may have evolved from?

    a.

    b.

15. Name two animals with finned feet and front flippers that are not closely related to each other.

    a.

    b.

16. *True or False*: Darwin recognized that the fossil record did not match what his theory predicted.

17. *True or False*: Since Darwin's time, millions of (difficult-to-fossilize) soft-bodied plants and animals have been found as fossils.

18. *True or False*: If evolution is true and if the fossil record is nearly complete, the fossil record should show one animal slowly changing into a completely different type of animal over time.

19. What would the fossil record show if the fossil record was nearly complete and if evolution was not true?

20. Give five examples of soft-bodied animals or plants that have been found as fossils.

    a.

    b.

    c.

    d.

    e.

21. *True or False*: An invertebrate is an animal with a backbone.
22. *True or False*: *T. rex* was the largest meat-eating dinosaur to have ever lived.
23. *True or False*: Cetaceans are a group of land mammals.
24. *True or False*: Different evolution scientists express different ideas concerning which land mammal became a whale.
25. *True or False*: A scientist can tell if an animal had a fluke by looking at the bones of the tail.
26. *True or False*: *Ambulocetus* was considered a whale only because evolution scientists believed it evolved into a whale, not because it had a blowhole or a whale's tail.
27. *True or False*: Some evolution scientists have put *Basilosaurus* on the evolutionary line to modern whales even though other evolution scientists believe this is not true.
28. *True or False*: According to scientists who support evolution, whale evolution is considered to be one of the best fossil proofs for evolution.
29. *True or False*: Scientists who support the idea that whales evolved from a land mammal cannot come to a consensus regarding which land mammal it was.
30. What land animal did Charles Darwin believe could have evolved into a whale?

31. There are two major problems with hippos being the evolutionary ancestor of whales. What are these two problems?

    a.

    b.

32. *True or False*: Superficially, *Archaeopteryx* looks like a modern bird.

33. The scientific name for flowering plants is ____________________.

34. According to evolution scientists, how long ago did life begin?

35. *True or False*: Organic chemicals have successfully been formed in the laboratory.

36. All living organisms today are comprised of these three necessary components:

37. ____________ contains the genetic blueprint of life.

38. *True or False*: The proposed earliest form of life, a single-celled bacteria-like organism, has never been (re)produced in a laboratory using chemicals.

39. Which is more likely, winning the National Powerball Lottery every day for 365 days in a row or DNA forming naturally to create life?

40. How many letters of DNA form in a natural strand of DNA?

41. How many letters of DNA would be needed to form just one protein 300 amino acids long?

42. What is the purpose of spiraling DNA?

43. *True or False*: Conceptually, proteins look like a chain.

Short Answer worth 7 points each

44. List three of the best evidences for evolution.

a.

b.

c.

45. List three of the best evidences for creation.

a.

b.

c.

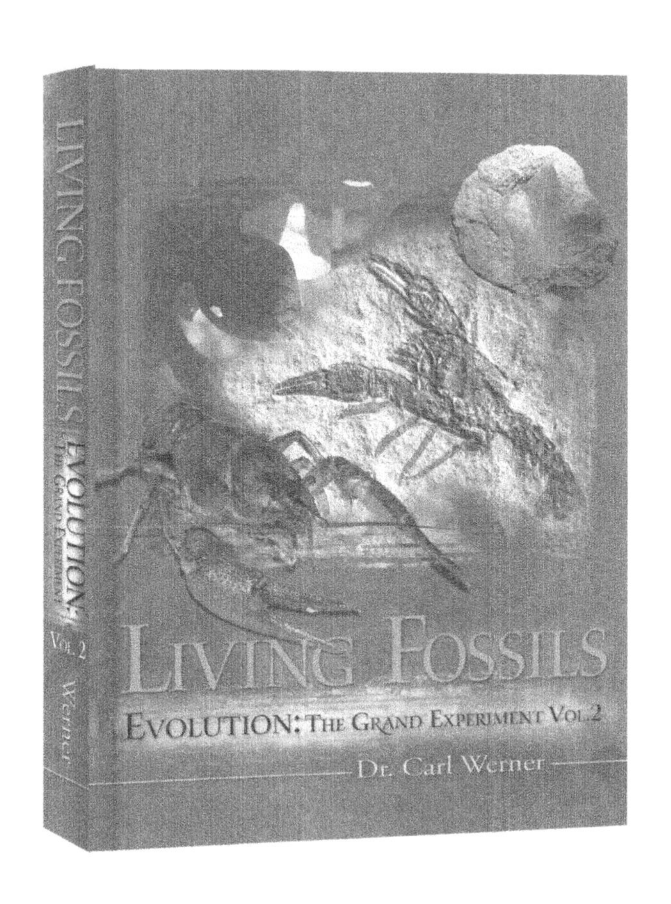

Sectional & Final Exams

for Use with

Living Fossils

Living Fossils	Sectional Exam 1	Scope: Chapters 1–11	Total score: ____of 100	Name
Concepts & Comprehension				

1-10. Outline Darwin's theory of evolution using only 10 words.

a.

b.

c.

d.

e.

f.

g.

h.

i.

j.

11. What year did Darwin publish the *The Origin of Species*?

12-14. What three errors did Haeckel commit to falsely prove evolution using the idea of Ontogeny Recapitulates Phylogeny?

a.

b.

c.

15–18. Name the four most well-recognized examples of living fossils from the dinosaur layers.

a.

b.

c.

d.

19-20. Name two scientific ideas that were believed by scientists for years but were later shown to be false.

a.

b.

21-23. Name the three evidences against evolution that caused the author to question and test the theory of evolution.

a.

b.

c.

24. What is the definition of a "living fossil"?

25. *True or False*: The fossil magnolia leaf found in dinosaur layers looks very similar to a modern magnolia leaf but was assigned a different species name.

26. *True or False*: Species names assigned may reflect the bias of the scientist who discovered the fossil.

27. *True or False*: A species is defined as two animals that can produce offspring.

28. *True or False*: From biggest group to smallest, the correct order is: kingdom, phylum, class, order, family, genus, and species.

29. *True or False*: If evolution is not true, then animals would not have changed dramatically over time.

30-31. List the two types of crinoids.

a.

b.

32-36. Name five classes of echinoderms that have been found in dinosaur rock layers that appear similar to modern forms.

a.

b.

c.

d.

e.

37-38. List the two types of sea urchins found in dinosaur rock layers that look modern.

a.

b.

39. *True or False*: Echinoderms have 5-fold symmetry.

40-44. Name five major types of aquatic arthropods living today that have been found in dinosaur layers.

a.

b.

c.

d.

e.

45–46. Name the bird and dinosaur found in Solnhofen, Germany, along with modern-appearing aquatic arthropods.

a.

b.

47. Name the phylum that includes starfish.

48. What percentage of the living land Arachnid orders have also been found in dinosaur rock layers?

49. *True or False*: Over time, the author began to trust less in the species names assigned to the fossils and relied more on his own judgement by comparing photographs of the fossils with the living organisms.

50-53. Name four common bivalve shellfish living today that were also found in dinosaur rock layers.

a.

b.

c.

d.

54. *True or False*: Tusk shells look like elephant tusks and are just as large.

55-59. Match these five classes from the Phylum Mollusca to their common names.

55. Bivalves

56. Snails

57. Chambered Shellfish

58. Tusk Shells

59. Sea Cradles

a. Class Polyplacophora

b. Class Scaphoda

c. Class Gastropoda

d. Class Cephalopoda

e. Class Bivalvia

60. *True or False*: Modern-appearing shellfish from all five major classes of shellfish living today in the Phylum Mollusca have been found in dinosaur rock layers.

61. *True or False*: Examples from both classes of segmented worms living today have also been found as fossils in dinosaur rock layers.

62-64. List the three classes of sponges living today. You may use formal names or informal names of classes.

a.

b.

c.

65. *True or False*: All three classes of sponges living today were also alive during the time of the dinosaurs.

	Living Fossils	Sectional Exam 2	Scope: Chapters 12-25	Total score: ____of 100	Name
	Concepts & Comprehension				

1-3. Name three modern types of bony fish found in dinosaur rock layers.

a.

b.

c.

4-6. Three cartilaginous fish were found in Solnhofen, Germany, along with a dinosaur and the famous toothed bird *Archaeopteryx*. What types of fish did these fossil fish look like?

a.

b.

c.

7-8. List the two types of jawless fish living today.

a.

b.

9. *True or False*: The dinosaur-era salamander called *Karaurus* was built like a modern salamander.

10. *True or False*: Fossil frogs have been found in Jurassic rock layers.

11. *True or False*: Frogs are members of the Phylum Arthropoda.

12. *True or False*: Crocodiles and alligators are members of the Phylum Chordata.

13-15. Name three types of crocodilians that are alive today that were also alive during the time of the dinosaurs.

a.

b.

c.

16. *True or False*: Snakes belong to the Phylum Chordata.

17-23. Modern-appearing animals from all of the seven major animal phyla have been found in dinosaur rock layers. What are these seven phyla?

a.

b.

c.

d.

e.

f.

g.

24-27. List four types of lizards living today that have been found in dinosaur rock layers.

a.

b.

c.

d.

28-29. The two lizard-like animals found in the same dinosaur rock layers as *Archaeopteryx* resemble which two modern reptiles?

a.

b.

30-33. Name the five groups of vertebrates living today that have also been found in dinosaur rock layers. Start the list with fish.

a. Fish

b.

c.

d.

e.

34-36. Name three birds that were alive during the time of the dinosaurs that went extinct.

a.

b.

c.

37. *True or False*: Mammals were found in dinosaur rock layers long before Darwin wrote *The Origin of Species.*

38-44. Match the animals on the left to the subclass of mammal on the right.

38. Opossum	a: Placental Mammal
39. Echidna	b: Marsupial Mammal
40. Tasmanian Devil	c: Monotreme Mammal
41. Kangaroo	
42. Duck-billed platypus	
43. Shrew	
44. Human	

45-47. List three mammals (either species name or type of mammal) that have been found in dinosaur layers that look like modern forms.

a.

b.

c.

48-50. Name three examples of cone-bearing plants alive today that have been found in dinosaur rock layers.

a.

b.

c.

51-53. Match the common and official names for these groups of spore-forming plants.

51. Moss	a. Division Pteridophyta
52. Club moss	b. Division Lycopodiophyta
53. Ferns and horsetails	c. Division Bryophyta

54-59. Name six types of flowering plants that are alive today that have been found in dinosaur rock layers.

a.

b.

c.

d.

e.

f.

60. *True or False*: All seven major plant divisions living today and all seven major animal phyla living today have been found in dinosaur rock layers, and many of these dinosaur-era organisms appear nearly the same as those living today.

61-65. Write out five major problems with the theories of the big bang and evolution.

a.

b.

c.

d.

e.

Living Fossils	Final Comprehensive Exam 1	Scope: Chapters 1–25	Total score: ____of 100	Name
Concepts & Comprehension				

1-10. Outline Darwin's theory of evolution using only 10 words.

a.

b.

c.

d.

e.

f.

g.

h.

i.

j.

11-13. What three errors did Haeckel commit to falsely prove evolution using the idea of Ontogeny Recapitulates Phylogeny?

a.

b.

c.

14–17. Name the four most well-recognized examples of living fossils from the dinosaur layers.

a.

b.

c.

d.

18-19. Name two scientific ideas that were believed by scientists for years but were later shown to be false.

a.

b.

20-22. Name the three evidences against evolution that caused the author to question and test the theory of evolution.

a.

b.

c.

23. What is the definition of a "living fossil"?

24. *True or False*: Species names assigned may reflect the bias of the scientist who discovered the fossil.

25. *True or False*: From biggest group to smallest, the correct order is: kingdom, phylum, class, order, family, genus, and species.

26-30. Name five classes of echinoderms that have been found in dinosaur rock layers that appear similar to modern forms.

a.

b.

c.

d.

e.

31-32. List the two types of sea urchins found in dinosaur rock layers that look modern.

a.

b.

33-37. Name five major types of aquatic arthropods living today that have been found in dinosaur layers.

a.

b.

c.

d.

e.

38–39. Name the bird and dinosaur found in Solnhofen, Germany, along with modern-appearing aquatic arthropods.

a.

b.

40. What percentage of the living land Arachnid orders have also been found in dinosaur rock layers?

41-44. Name four common bivalve shellfish living today that were also found in dinosaur rock layers.

a.

b.

c.

d.

45. *True or False*: Tusk shells look like elephant tusks and are just as large.

Match these five classes from the Phylum Mollusca to their common names.

46. Bivalves
47. Snails
48. Chambered Shellfish
49. Tusk Shells
50. Sea Cradles

a. Class Polyplacophora
b. Class Scaphoda
c. Class Gastropoda
d. Class Cephalopoda
e. Class Bivalvia

51-53. Name three modern types of bony fish found in dinosaur rock layers.

a.

b.

c.

54-56. Three cartilaginous fish were found in Solnhofen, Germany, along with a dinosaur and the famous toothed bird *Archaeopteryx*. What types of fish did these fossil fish look like?

a.

b.

c.

57-59. Name three types of crocodilians that are alive today that were also alive during the time of the dinosaurs.

a.

b.

c.

60-66. Modern-appearing animals from all of the seven major animal phyla have been found in dinosaur rock layers. What are these seven phyla?

a.

b.

c.

d.

e.

f.

g.

67-70. List four types of lizards living today that have been found in dinosaur rock layers.

a.

b.

c.

d.

71-74. Name the five groups of vertebrates living today that have also been found in dinosaur rock layers. Start the list with fish.

a. Fish

b.

c.

d.

e.

75-77. Name three birds that were alive during the time of the dinosaurs that went extinct.

a.

b.

c.

78. *True or False*: Mammals were found in dinosaur rock layers long before Darwin wrote *The Origin of Species*.

79-85. Match the animals on the left to the subclass of mammal on the right.

79. Opossum
80. Echidna
81. Tasmanian Devil
82. Kangaroo
83. Duck-billed platypus
84. Shrew
85. Human

a. Placental Mammal
b. Marsupial Mammal
c. Monotreme Mammal

86-88. List three mammals (either species name or type of mammal) that have been found in dinosaur layers that look like modern forms.

a.

b.

c.

89-91. Name three examples of cone-bearing plants alive today that have been found in dinosaur rock layers.

a.

b.

c.

92-94. Match the common and official names for these groups of spore-forming plants.

92. Moss
93. Club moss
94. Ferns and horsetails

a. Division Pteridophyta
b. Division Lycopodiophyta
c. Division Bryophyta

95. *True or False*: All seven major plant divisions living today and all seven major animal phyla living today have been found in dinosaur rock layers, and many of these dinosaur-era organisms appear nearly the same as those living today.

96-100. Write out five major problems with the theories of the big bang and evolution.

a.

b.

c.

d.

e.

Answer Keys for Worksheets

for Use with

Evolution: The Grand Experiment

Chapter 1

The Origin of Life: Two Opposing Views

1. a. An all-powerful God created the entire universe and all forms of life, such as humans and dinosaurs, at the same time (creation).
   b. The universe began naturally, billions of years ago as a result of an explosion (big bang). Later, a bacterium-like organism arose spontaneously from a mixture of chemicals (abiogenesis) and this single-cell organism evolved into all modern life forms, including humans (evolution).
   c. A third view is that God caused the big bang and then helped the process of evolution along (big bang and evolution with God's aid).
2. The theory that the universe was created by a large explosion in space 10 to 20 billion years ago. The theory is based on the observation of the universe expanding.
3. 1859
4. a. Scientists have collected nearly one billion fossils.[1]
   b. They have described the structure and function of DNA.
   c. They have identified how genes are passed on to the next generation.
5. a. Michelangelo
   b. The ceiling of the Sistine Chapel
   c. Depicts God creating man
6. Scientists
7. a. Gaps in the fossil record
   b. Problems with the big bang theory
   c. The amazing complexity of even the simplest organisms
   d. The inability of scientists to explain the origin of life using natural laws
8. a. Observations of natural selection in action
   b. The evolution of birds from dinosaurs
   c. The evolution of whales from a land animal
   d. The evolution of man from apes
9. Some fear that teaching two opposing theories would confuse the students. Others believe that creation is a religious idea and should not be taught in government schools.
10. Some believe this approach would encourage students to think critically and openly about the world around them.
11. a. According to the Gallup poll presented in this chapter, parents want creationism taught in public schools along with evolution so that students can learn the facts and evidences both for and against each theory.
    b. Yes-54%
    c. No-22%
    d. Unsure-24%

Chapter 2

Evolution's False Start: Spontaneous Generation

1. a. The theory of spontaneous generation
   b. Fourth century B.C.
   c. Over 2100 years
   d. 1859
2. Living organisms could come into being rapidly and "spontaneously" over a period of just a few days or weeks.
3. "If you press a piece of underwear soiled with sweat, together with some wheat in an openmouthed jar, after about 21 days the odor changes and the fermentation, coming out of the underwear and penetrating through the husks of wheat, changes the wheat into mice...But what is even more remarkable is that the mice, which come out of the wheat and underwear, are not small mice, not even miniature adults or aborted mice, but adult mice emerge!"
4. Questioning spontaneous generation was tantamount to questioning science itself. They would be thought of as fools.
5. Put a piece of meat into an open jar. Wait two weeks for spontaneous generation of maggots on the meat.
6. a. When Dr. Redi placed a piece of cheesecloth over the jar of meat, maggots never formed because flies were unable to land on the meat and lay their eggs.

---

1 Older editions of *Evolution: The Grand Experiment* may have 200 million or "hundreds of millions" as the number of collected fossils. See footnote 1 in Appendix A of *Evolution: The Grand Experiment.*

b. It was proof against the theory of spontaneous generation.

7. 1668

8. They took clear pond water, boiled it, and poured it into a jar. After a few weeks or so, the pond water became cloudy and scum formed on the water.

9. He boiled broth to kill the living things that may have been present in the liquid but later the broth turned cloudy.

10. a. Dr. Pasteur started with a flask filled with boiled meat broth. He then heated the neck of the flask and stretched it. After heating the neck of the flask again, Pasteur then made it s-shaped. The open end of the s-shaped glass neck pointed upward. Due to gravity, bacteria from the air could only settle in the lowest part of the neck and were prevented from reaching the broth. Even after months of waiting, the liquid in the flask never became cloudy. Pasteur then tilted the flask, allowing the liquid in the flask to come in contact with the neck. Within a short period of time, the liquid became cloudy. The microorganisms that had settled in the neck contaminated the broth.

    b. This proved that bacteria (or in this case scum) do not form spontaneously from clear water!

11. 1859

12. The theory that all life must come from pre-existing life.

13. Scientists can be wrong, even though they may be confident in their convictions.

14. A generous dose of skepticism goes a long way in science. Also, a scientific idea may not be disproved for hundreds, if not thousands, of years.

Chapter 3

Darwin's False Mechanism for Evolution: Acquired Characteristics

1. 1859

2. *On the Origin of Species by Means of Natural Selection, or the Preservation of Favoured Races in the Struggle for Life*

3. A primordial prototype

4. A primordial single-cell organism

5. a. A single-cell organism

    b. Evolved into a multicellular invertebrate

    c. Which evolved into a fish

    d. Evolved into a semi-aquatic amphibian

    e. Evolved into a land-based reptile

    f. One type of land-based reptile changed into a bird

6. a. A single-cell organism

    b. Evolved into a multicellular invertebrate

    c. Which evolved into a fish

    d. Evolved into a semi-aquatic amphibian

    e. Evolved into a land-based reptile

    f. Another type of land-based reptile changed into a mammal

    g. Evolved into a primate (ape)

    h. Evolved into humans

7. a. Millions and millions of years

    b. It supposedly took 2-3 weeks for spontaneous generation to generate life.

8. The disproved idea that changes acquired in the body during life, such as enlarged muscles or a suntan, can be passed on to the next generation.

9. a. The law of use and disuse

    b. "Lamarckianism"

10. a. "I think that there can be little doubt that use in our domestic animals strengthens and enlarges certain parts, and disuse diminishes them; and that such modifications are inherited."

    b. *The Origin of Species*

11. Body cells have no influence on the DNA in the reproductive cells (eggs and sperm). It is only the genes in the reproductive cells that are passed on to the next generation.

12. a. Muscle building- even though a man lifts weights every day and develops large muscles, his baby will not be born with large muscles.

    b. Neck stretching- Stretching neck muscles has no effect on the DNA in the reproductive cells of the horse. A longer neck cannot be passed on to the next generation.

    c. Sun tanning- If a woman tans her skin every day, her children will not be born with a suntan. Tanning has no effect on the genes of the reproductive cells; therefore, tanned skin is not passed on to the next generation.

d. Disuse and shedding body parts- The DNA in the reproductive cells of a multicellular animal cannot sense that another part of the body is not being used. Therefore, an animal's offspring are not affected by the disuse of a body part.

13. a. "If this implies that many parts are not modified by use and disuse during the life of the individual, I differ widely from you, as every year I come to attribute more and more to such agency."

    b. 1875

    c. He wrote this to his cousin, Francis Galton.

14. Scientist, August Weisman, disproved the concept of disuse. He cut off the tails of mice for 20 generations in a row. No matter how many tails he cut off, the baby mice were always born with a tail.

15. 1889

Chapter 4

Natural Selection and Chance Mutations

1. a. Artificial breeding

   b. Natural selection

2. Answers should include some of the following ideas:

   Natural selection, or the killing off of weaker varieties of animals or plants within a species, only removes these weaker animals and plants, but it does not add completely new body parts to the population.

   A particular animal or plant species may have five varieties of color or shape or size coded in its genes. Using artificial breeding, a horticulturist or a breeder consciously (or artificially) selects which of these traits will be eliminated or preserved in order to create a new variety (or breed) of plant or animal within a species.

   Natural selection and artificial breeding may remove a trait that is already present; however, these breeding processes never create new information in the DNA and never create new genes. They simply select traits that are already present.

3. Nature and the environment acting as the selectors may remove a particular trait by killing off all of the animals with that trait. For instance, in the snowy environment of the Arctic Circle, if there were two varieties of a particular bear species, one with white fur and one with dark gray fur, the white variety would tend to survive better than the dark gray ones. This is because the snowy background prevents the white bears from being easily spotted when hunting their prey. Since the dark gray bears do not have the advantage of a lighter color, this variety of bear could not as easily sneak up on its prey. By a natural process, the darker bear would eventually be eliminated as it is less able to catch its prey.

4. Within any particular animal or plant species, there are known limits in size, limits in varieties of color, and limits in shape, as well as limits in other characteristics.

5. Dr. Thomas Hunt Morgan

6. a. 1910

   b. Columbia University

7. Dr. Morgan observed that fruit flies were normally born with red eye color and only rarely, once in every 100 generations or more, were they born with white eyes. Since the white eye color was not a normal trait, he correctly deduced that the albino eye occurred through an accident or mutation.

8. a. Sickle-cell anemia

   b. Cystic fibrosis

   c. Spina bifida

   d. Hemophilia

9. Thousands

10. a. To form a complicated new body system would require adding or changing thousands of letters of DNA in the egg or sperm of the parent organism, not just one letter. Moreover, these thousands of letters of DNA would not only have to be accidentally placed in the correct location, but also in the correct letter sequence.

    b. Producing a cardiovascular system from blind mutations in the DNA code would be much more complex than typing "Pick up Mary after dance lessons." Rather it would be equivalent to ten million blindfolded children authoring a short book.

11. The use of the word "adaptation" has become a

matter of semantics and is somewhat confusing to the non-scientist. When a modern evolution scientist says an animal has "adapted," he or she actually means that accidental mutations have occurred, resulting in an animal which is better suited for the environment. Many incorrectly believe that when an animal "adapts," the animal changed in response to the environment and has done so out of necessity, purposefully.

12. Although the public's perception of adaptation provides a more believable mechanism for evolution, it is incorrect and is very different from how a scientist uses the term.

13. Student should summarize 3 of the 6 examples.

    Example #1: Hunting larger game nearly doubled the size of man's brain, transforming one breed of *Australopithecus* to *Homo erectus*. Incorrect. Hunting larger animals does not affect the DNA of reproductive cells, and would not cause an ape to change into a human being. These changes could only theoretically come about by mistaken mutations and natural selection.

    Example #2: Around five million years ago upright walking became the primary form of locomotion in the Australopithecines due to the severe drought to the east of the African Rift Valley. Misleading. An animal cannot change from walking on all fours to upright walking because of a drought or a change in the environment. Environmental changes have no effect on the DNA of reproductive cells.

    Example #3: "The ape's massive jaw may have evolved as an adaptation to a diet of tough meat, raw or lightly cooked meat." Incorrect. Eating meat will not cause a change in the jaw size that can be passed to the next generation. Eating specific types of food has no effect on the DNA of reproductive cells.

    Example #4: "A whole suite of human characteristics indicate very specific adaptation to an environment of high radiant heat stress." Misleading. An individual animal cannot directly change the DNA in its reproductive cells because of a change in the environment.

    Example #5: Tool-using and tool-making have been important catalysts in human evolution. Genetically impossible. Using a tool or making a tool does not affect the DNA of reproductive cells.

    Example #6: The diversity of environments that humans have colonized during the last 100,000 years helps to explain their physical differences. Not true. Racial differences are not brought about by an environmental change, but are simply variations of the DNA in the genes.

14. a. Whales
    b. Dolphins
    c. Porpoises

15. They range in size from the tiny finless porpoise measuring 1.2 meters to the enormous blue whale measuring 100 feet long.

16. a. 100 feet long
    b. 400,000 pounds
    c. About the weight of a full-grown elephant
    d. The blood vessels are so large a human could swim through them.
    e. The size of a small car
    f. Yes

17. Because whales are warm-blooded mammals, Darwin reasoned that whales evolved from another warm-blooded mammal, possibly a bear, by means of acquired characteristics and natural selection. He thought that if a bear swam in the water for hours trying to catch insects, eventually it could be changed into a whale.

18. Professor Richard Owen, Director of the British Museum of Natural History

19. Answers should include three of these five animals:
    A: A hyenalike animal called Pachyaena
    B: A hippo-like animal
    C: A cat-like animal called Sinonyx
    D: A wolf-like land animal
    E: A deer-like animal called Indohyus—similar in appearance to the modern mouse deer

20. Whales

21. Preposterous

22. a. The hyena would have to develop a dorsal fin.
    b. The bony tail of the hyena would have to change into a cartilaginous fluke.
    c. The hyena's teeth would have to develop into a huge baleen filter.
    d. The hyena's hair would have to nearly

disappear and be replaced by blubber for insulation through chance mutations in the DNA.

e. The nostrils would have to move from the tip of the hyena's nose to the top of the whale's head, disconnect from the mouth passage, and form a strong muscular flap to close the blowhole.

f. The hyena's front legs would have to change into pectoral fins.

g. The hyena's body would have to increase in size from 150 pounds to 400,000 pounds.

h. The hyena's external ears would have to disappear and then develop to compensate for high-pressure diving to 1,640-feet deep.

i. The hyena's back legs would have to disappear.

23. # of body changes x # of new proteins needed per change x # of amino acids in each new protein x 3 new letters of DNA needed for each new amino acid added = Total # of DNA letters that must be added

24. a. A
    b. C
    c. G
    d. T

25. a. 1 out of 4 possibilities or ¼
    b. ¼ times ¼
    c. ¼ times ¼ times ¼

26. ¼ times ¼ times ¼ times and so on, 2,700 times, or ¼ 2,700 (Change to ¼ to the 2700 power)

27. a. The odds of one individual winning the Powerball Lottery once every year, for 200 years in a row, would be 1/1,149 followed by 1,577 zeros.
    b. Winning the Powerball Lottery once every year for 200 years in a row.

Chapter 5

Similarities: A Basic Proof of Evolution?

1. The term "related animals" theoretically means that two or more animals evolved from a common ancestor and are thus closely related to one another to the extent that they inherited similar body features, such as the same number of bones in the arm or the same number of fingers in the hand.
2. "Unrelated animals" theoretically means that two or more animals are only distantly related to one another since they do not share a common ancestor that has similar body features.
3. According to the theory of evolution, all four of these animals evolved from a theoretical common ancestor that also had one bone in the upper arm, two bones in the lower arm, and multiple bones in the wrist/hand/fin regions.
4. One is a fish and one is a mammal, and neither evolved from a theoretical common ancestor with a dorsal fin.
5. a. Both have an extra thumb on their hands.
   b. Both have a V-shaped jaw.
   c. Both have similar teeth.
   d. Both have similar skulls.
6. Scientists tested the DNA and found the giant panda belongs to the bear family and the red panda is related to raccoons.
7. a. Both have front flippers.
   b. Both have finned feet.
8. a. Seals - skunk or otter
   b. Sea Lions- a dog or bear
   c. They do not share a common ancestor.
9. Rhinoceros and horse
10. An elephant and sea cow
11. According to evolution, it is the phenomena where two groups of organisms, possibly distantly related, evolve into similar patterns and come to look like one another.
12. They all have a dorsal fin, a pectoral fin, and a finned tail.
13. In this case, scientists ignored similarities in the dorsal fins and instead focused on other features such as fur, mammary glands, and the type of teeth.
14. a. Pterosaur
    b. Bat
    c. Bird
15. a. Nautilus (mollusk)
    b. Human (vertebrate)
    c. Hoverfly (arthropod)

16. a. Duck-billed platypus (mammal)
    b. Duck-billed dinosaur (reptile)
    c. Duck-billed bird (bird)
17. a. Penguin (bird)
    b. Ichthyosaur (reptile)
    c. Bluefin tuna (fish)
18. a. Cassowary (bird)
    b. Oviraptor (reptile)
19. a. Placental mouse and marsupial mouse
    b. Placental mole and marsupial mole
20. Whale
21. Horse
22. Each investigator decides on the basis of an assumed evolutionary pathway which he or she prefers, and then interprets the evidence accordingly, invoking whatever reversals, parallel evolution, or convergent evolution, as the scheme may require. Such theories are so plastic that they are rendered non-falsifiable even if false, and thus they cannot be called scientific theories.

Chapter 6

The Fossil Record and Darwin's Prediction

1. The fossil record is the documented collection of animal and plant fossils known worldwide.
2. The fossils collected by scientists prior to 1859 did not correspond with his theory of evolution.
3. "Why then is not every geological formation and every stratum full of such intermediate links? Geology assuredly does not reveal any such finely-graduated organic chain; and this, perhaps, is the most obvious and serious objection which can be urged against the theory."
4. He argued that natural selection is correct despite the fact that the fossils don't particularly support it.
5. The fossil record would reflect many of the intermediate or ancestral forms of an animal slowly changing into a different type of animal over time.
6. There would be no intermediate animals.
7. Close to one billion[2]
8. a. The number of fossil fish in museums: 500,000
   b. The number of fossil dinosaurs in museums: 100,000
   c. The number of fossil bats in museums: 1,000
   d. The number of fossil bird specimens in museums: 200,000+
   e. The number of fossil pterosaurs (flying reptiles) in museums: nearly 1,000
   f. The number of fossil insects in museums: 1,000,000
   g. The number of fossil plants in museums: 1,000,000
   h. The number of fossil turtles in museums: 100,000
   i. The number of fossil whales: 4,000
   j. The number of fossilized soft-bodied animals, plants, and other soft-bodied organisms that have been found: millions
9. Bacteria, plants, leaves, embryos, worms, cattails, fish eggs, jellyfish, flowers
10. a. Over 750,000,000
    b. Over one million
11. The fossil record is comprehensive, balanced, accurate, and impressive.
12. 97.7%
13. 79.1%
14. 87.8%
15. Answers may vary but students should show an understanding of the completeness of the fossil record and give an informed opinion based on sound reason and facts.

Chapter 7

The Fossil Record of Invertebrates

1. Animals without a backbone or spinal cord
2. Animals with a backbone
3. More than 200,000
4. None
5. Sea pen
6. They appear suddenly in the Cambrian fossil layers without a trace of an evolutionary ancestor. Yet over 15,000 species of trilobites have been

2 Older editions of *Evolution: The Grand Experiment* may have 200 million or "hundreds of millions" as the number of collected fossils. See footnote 1 in Appendix A of *Evolution: The Grand Experiment.*

collected. If evolution occurred and if the fossil record is reflective of the past, then the animals that evolved into trilobites should have been discovered by now. Hundreds of thousands of fossil trilobites, possibly millions, have been collected by museums, but no direct ancestors have been found.

7. More than 750,000,000
8. The sudden appearance of the phyla groups in the lower Cambrian/Ediacaran fossil layers.
9. No, Darwin expected there to be fossil ancestors found showing evolution. He expected the fossil record to show this as more fossils were found. The millions of fossils found do not support Darwin's predictions.
10. Scientists who support evolution suggest ancestors of the invertebrate phyla groups existed but were soft-bodied animals and less likely to fossilize; this is why these invertebrate groups appear without ancestors.
11. Scientists who oppose evolution point out that thousands of fossils of soft-bodied animals (jellyfish, worms, sponges, soft corals, etc.) have been found in the lowest fossil layers (Cambrian/Ediacaran) plus fossils of microscopic bacteria. They cite the absence of theoretical evolutionary ancestors of the invertebrate phyla groups—in light of a vast fossil record, including fossils of soft-bodied animals (and bacteria)—as evidence that evolution did not occur.
12. a. Devonian
    b. Silurian
    c. Ordovician
    d. Cambrian
    e. Ediacaran
13. Large numbers
14. None
15. Students' answers may vary but they should give sound logic and examples for their conclusions.

Chapter 8

The Fossil Record of Fish

1. 500,000
2. Many are exquisitely preserved with all of the bones, scales, and fins readily apparent.
3. More than 750,000,000
4. a. No
   b. No
   c. Fish families appear suddenly without fossilized transitional ancestors.
5. None
6. They suggest that if the evolution truly occurred, one should be able to witness the transformation from invertebrates to vertebrate fish. One should also see one fish family slowly changing into another.
7. "...the transition from spineless invertebrates to the first backboned fishes is still shrouded in mystery, and many theories abound as to how the changes took place. There are still many different opinions as to which invertebrate group may have given rise to the first vertebrates or first fishes...I'm sure that in the next 10 years or so we'll answer this mystery."
8. It's a mystery.
9. It's a mystery.
10. The fossils tell us little about which other group of fishes they may have evolved from or collaterally with. It's a mystery.
11. It's a mystery.
12. It is still very much debated amongst paleontologists. It's a mystery.
13. Answers may vary but the student should give examples to support their conclusions.

Chapter 9

The Fossil Record of Bats

1. Over 1,000
2. One would expect to find fossil ancestors showing the evolution of the bat, such as the formation of a bat's wing membrane over time or the slow elongation of the mammal's fingers to support a wing membrane.
3. Answers may vary but should include the idea that bats would be fully formed and found suddenly in the fossil record.
4. None
5. No
6. Since fossilization is a chance event and rarely occurs, the fossil record can, at times, be

incomplete. Therefore, the absence of bat ancestors does not necessarily imply they did not exist.

7. If the ancestors existed, surely some of them should have been fossilized. They believe it is inconsistent to blame the lack of bat ancestors on the processes of fossilization since millions of other fossils from the time-period of bat evolution have been fossilized and collected by museums. They suggest this pattern of absence of ancestors is strong evidence that evolution did not occur.
8. Eocene
9. They are the same.
10. No
11. "We have no fossil records of bats during the Cretaceous period. This means that we are only depending on speculation, when it [bat evolution] started and what happened in that time."

Chapter 10

The Fossil Record of Pinnipeds: Seals and Sea Lions

1. Sea lions have a visible external ear.
2. a. California sea lion
   b. Australian sea lion
   c. Stellar sea lion
   d. Up to 25 miles per hour
3. a. A dog-like animal
   b. A bear-like animal
4. No
5. None
6. They hold onto hope that one day these fossils will be discovered, given enough time and money to search for these fossils.
7. a. Pithanotaria
   b. It's very similar in terms of its body size and morphology to the modern sea lions.
8. Thousands
9. B-0; C-0; D-0; E-0; F-0; G-0; H-0
10. a. Nearly a mile (around 5,200 feet deep)
    b. A Los Angeles class nuclear attack submarine can dive only 800 feet deep.
    c. California sea lions can readily dive up to 360 feet deep and occasionally dive to 800 feet.
11. Seals can hold their breath for an amazing two hours at a time. Sea lions can hold their breath for up to 15 minutes.
12. a. A skunk-like animal
    b. An otter-like animal
13. No
14. None
15. 5,000
16. The fossil record of seals does not match Darwin's predictions.
17. B-0; C-0; D-0; E-0; F-0; G-0; H-0; I-5,000

Chapter 11

The Fossil Record of Flying Reptiles

1. A reptile
2. Dinosaurs
3. Nearly 1,000 have been found on every continent, including Antarctica.
4. Some of these fossils are exquisitely preserved with detailed impressions of the soft membranes of the wing.
5. a. Pterodactyls have short tails.
   b. Rhamphorhynchus have long tails.
6. a. Pterosaurs range in size from that of a tiny sparrow to larger than a fighter jet.
   b. The wingspan of the largest known pterosaur was two feet wider than that of a U.S. F-4E Phantom II fighter jet.
7. It's a mystery.
8. None
9. None
10. B-0; C-0; D-0; E-0; F-0; G-0; H-0; I-nearly 1,000
11. No

Chapter 12

The Fossil Record of Dinosaurs

1. *Tyrannosaurus rex*
2. 32
3. None
4. B-0; C-0; D-0; E-0; F-0; G-0; H-0; I-32
5. No

6. When scientists cannot find evidence that a dinosaur is the direct ancestor of another, they label these dinosaurs as cousins.
7. a. Cretaceous
   b. Jurassic
   c. Triassic
8. *Triceratops*
9. a. Ornithischians
   b. Saurischians
10. Ornithischians
11. Hundreds
12. None
13. A-0; B-0; C-0; D-0; E-0; F-0; G-0; H-0; I-hundreds
14. No
15. Brontosaurus
16. Sauropod
17. Nearly 30
18. None
19. A-0; B-0; C-0; D-0; E-0; F-0; G-0; H-0; I-30
20. No
21. Over 30 million
22. Thousands
23. Over 700
24. No
25. None

Chapter 13

The Fossil Record of Whales Worksheet 1

1. Whales are warm-blooded, give birth to live young, suckle their young, and have some body hair around their face used for tactile sensation. Based on these characteristics, whales are classified as mammals.
2. a. Whales
   b. Dolphins
   c. Porpoises
3. Reptiles
4. One species of land mammal went back into the water and evolved into a whale 50 million years ago.
5. The fossil evidence for the evolution of whales is considered both strong and unique by scientists who support evolution. Scientists who oppose evolution believe the fossil evidence demonstrating whale evolution is simply wishful thinking on the part of evolution scientists.
6. It is generally acknowledged as rare to have fossils showing one animal group slowly evolving into a completely different animal.
7. "The paleontology of early whales is one of our most widely and justly trumpeted success stories."
8. "We now have whales with legs, whales with reduced legs, whales with little tiny legs, whales with no legs at all, and their heads are getting bigger and their teeth are getting stranger...You really have to be blind or three days dead not to see the transition among these. You have to not want to see it."
9. "…They do show us intermediates in the evolution of whales. We don't often get fossil intermediates so we can actually trace the development of characteristics, say, for example, the evolution of swimming in whales. We don't often have that opportunity."
10. a. *Sinonyx jiashanensis*
    b. *Ambulocetus natans*
    c. *Rodhocetus kasrani*
    d. *Dorudon atrox*
    e. *Basilosaurus isis*
11. Flawed
12. a. A hyena-like animal
    b. A cat-like animal
    c. A hippopotamus relative
13. Bears
14. He sensed he was losing a public relations battle. Professor Richard Owen, Director of the British Museum of Natural History, prevailed on Darwin to leave out the story or at least tone it down.
15. They say it indicates the story of whale evolution is just that, a story. They ask: Why can't the supporters of the theory of evolution agree on the ancestor of whales, given the "fact" that whale evolution is so clear to them using the fossils?
16. All whales are meat-eaters. Even large filter-feeding baleen whales eat small crustacean animals called krill. Evolution scientists have chosen certain meat-eating land mammals, because of the similarities of

their meat-eating teeth when compared to the teeth of the oldest fossil whales.

17. a. Scientists who oppose evolution point out that just because an animal has similar features does not necessarily indicate they are evolutionary ancestors to one another. Many animals that are not related have nearly identical body plans, such as the marsupial mole and the placental mole, as discussed in Chapter 5.

    b. DNA evidence suggests a *plant*-eating hippo-like mammal evolved into a meat-eating whale. Fossil evidence suggests it was a *meat*-eating cat-like or hyena-like mammal.

18. All whales are meat-eaters. Even large filter-feeding baleen whales eat small crustacean animals called krill.

19. Scientists at the Tokyo Institute of Technology have recently found evidence that hippopotamus DNA is the closest match to the DNA of whales (when compared to all of the other mammal groups).

The Fossil Record of Whales Worksheet 2

1. The whale has similarities to other animals that are quite different from each other. The same is true for the hyrax.

2. They do not believe that any of these animals shared a common ancestor. They propose that the similarities are either coincidental or the result of intentional design, as in the previous examples. For them, the whole process of comparing similarities is highly subjective and open to observer bias.

3. "We still have the problem, if we are talking about whales evolving from these even-toed hoofed mammals [hippos], they are all plant-eaters. Whales today are all carnivores."

4. Because whales have been around in the fossil record about five times as long as hippos have

5. Because *Ambulocetus* could both walk on land and swim and had whale-like, meat-eating teeth

6. *Ambulocetus* was defined as a "walking whale" not because it had a whale's tail or a whale's flippers or a blowhole, but because (some) evolution scientists believed it evolved into a whale. Once evolution scientists believed it was on the line to becoming a whale, it became a "whale." And since it was a land animal with four legs, it was then called a "walking whale." Scientists who oppose evolution are quick to point out that this reasoning is circular and therefore specious.

7. *Ambulocetus* had eyes on the top of its head, more like a crocodile.

8. A whale's tail

9. It had arms and legs but it could also swim like a whale.

10. By looking at the bones of the tail. A whale's tail has a special round "ball" vertebrae, followed by several flat bones where the cartilaginous fluke tail attaches.

11. He speculated that it might have had a fluke.

12. a. The ball vertebrae followed by several flat bones where the cartilaginous fluke tail attaches

    b. Hand bones

    c. Feet bones

13. *Basilosaurus*

14. Four

15. If whale evolution is the *best* example of evolution, then for all practical purposes, the theory is dead — especially in light of the fact that over two million fossil whale bones have been discovered representing thousands of whales.

Chapter 14

The Fossil Record of Birds Part 1: *Archaeopteryx* Worksheet 1

1. The nineteenth century

2. Some scientists believed they had found the missing link proving the evolution of birds from dinosaurs. They consider the evolution of birds one of the three best fossil proofs for the theory of evolution.

3. a. The evolution of birds from dinosaurs

   b. The evolution of whales from a land mammal

   c. The evolution of men from ape

4. They believe *Archaeopteryx* is a hybrid animal, possessing traits similar to a dinosaur (dinosaur tail, scaly reptilian head, and claws) and a bird (feathers and wings).

5. Germany

6. Nine

7. a. Half dinosaur, half bird
   b. Just a bird
8. Scientists assumed the lack of feathers around the head of *Archaeopteryx* fossils meant the animal did not have feathers on its head during life. They took this interpretation one step further and concluded that this animal must have had scales on its head (even though they did not see scales in the fossils).
9. Since modern birds frequently fossilize without feather imprints around the head, then *Archaeopteryx* probably had feathers on its head too.
10. With this simple reinstallation of feathers, *Archaeopteryx* no longer looks reptilian, nor does it look like a hybrid animal. Rather, it looks more like a modern bird.
11. It had a scaly reptilian head.
12. They have a feathered head and look much more like a modern bird.
13. They have suggested that claws on its wings indicate *Archaeopteryx* was the progeny of meat-eating dinosaurs (also with claws), such as *Deinonychus*.
14. The wings do not necessarily link *Archaeopteryx* to meat-eating dinosaurs. They point out that other flying vertebrates also have claws on their wings.
15. a. Bats
    b. Pterosaurs
    c. Ostriches (hoatzins, and touracos)
16. a. Ostriches
    b. Hoatzins
    c. Touracos

The Fossil Record of Birds Part 1: *Archaeopteryx* Worksheet 2

1. The tails of meat-eating dinosaurs are 4 to 5 feet long and covered with scales while the tail of *Archaeopteryx* is 4 to 5 inches long and is covered with feathers. Also, the tail of the reconstructed models of *Archaeopteryx* look rather like a modern bird's tail.
2. Since modern birds have short tails, proponents of evolution believe it looks more like a dinosaur. Those who oppose evolution believe the differences between the tail and a dinosaur tail are so dramatic, you cannot consider them to be related.
3. Teeth are a unique feature not seen in any modern bird.
4. Meat-eating dinosaurs have teeth that are serrated, like a steak knife, but the teeth of *Archaeopteryx* are smooth.
5. a. An ostrich is not related to a giraffe because it has a long neck.
   b. A duck is not related to the duck-billed platypus because it has a billed beak.
   c. A puffin bird is not related to a fish because it can swim deep in the water.
6. Only a whole series of fossils, showing a dinosaur slowly changing into a flying bird, such as *Archaeopteryx*, would prove the evolution of birds.
7. When a newer *Archaeopteryx* model is placed alongside modern birds, it does not appear to be very different.
8. Answers may vary but should be based on observations and comparisons given.
9. No
10. *Deinonychus* lived 30 million years after *Archaeopteryx* and, therefore, could not be the ancestor of birds.
11. a. *Deinonychus* is dated at 93 to 119 MYA while *Archaeopteryx* is dated at 150 MYA
    b. *Archaeopteryx* is older.
    c. Deinonychus lived 30 million years after Archaeopteryx and, therefore, could not be the ancestor of birds.
12. a. Deinonychus
    b. Triassic archosaurs
13. Because evolution scientists are not sure what type of animal evolved into birds. They ask: What does this say for the theory of evolution, as a whole, if bird evolution is touted as one of the three best fossil proofs for evolution? Given the extraordinarily rich bird and dinosaur fossil records, they argue that the ancestral gaps between reptiles and *Archaeopteryx* are too great.
14. 100,000 fossil dinosaurs and 200,000 fossil bird specimens have been collected, yet evolution science cannot demonstrate a single reptile (dinosaur or archosaur) evolving into a bird.
15. Any 5: sharks, guitar fish, horseshoe crabs, dragonflies, turtles, lizards, lobsters, crayfish,

shrimp, cockroaches, woodwasps, waterbugs, grasshoppers, beetles, scorpion flies, water skeeters, sea urchins, sea stars, and prawns.

16. They call them "living fossils" and say these types of animals were so well-adapted to the environment that they did not need to change and have therefore remained the same for over a hundred million years.
17. They believe that these fossils indicate that life has not evolved, simply that some animals, such as *Archaeopteryx* and dinosaurs, have gone extinct.

Chapter 15

The Fossil Record of Birds Part 2: Feathered Dinosaurs Worksheet 1

1. Mid 1990s
2. Liaoning Province
3. The missing links between dinosaurs and modern birds
4. The anatomy of the feathers and the size of the wings
5. Some of these specimens have teeth and some have other dinosaur-like features, such as a long tail.
6. The Chinese specimens are small, generally about the size of a modern bird, with most less than 12 inches long.
7. They are asymmetrical, meaning the quill (rachis) does not run down the center of the feather. It is instead off-centered.
8. They have *symmetrical* feathers; that is, the quill (rachis) does run down the center of the feather. They are less dense, less organized, and do not have barbules.
9. Modern flightless birds: the quill runs down the center of the feather making the feather symmetrical, the barbs are loose and less dense.
10. The rock layers containing the "feathered dinosaurs" are younger in age than the rock layers where the bird *Archaeopteryx* was found.
11. They ask: How could the "feathered dinosaurs" be the ancestors of birds if they lived after *Archaeopteryx*, a bird that could already fly?
12. Many of these Chinese fossils had been altered.
13. Although the fossil looks real, the white paint on the upper edge of the fossil indicates that repairs were made and the fossil is a composite.
14. It was painted.
15. a. The jaw bone
    b. There's a little piece of the radius [wing bone] that's upside down.

The Fossil Record of Birds Part 2: Feathered Dinosaurs Worksheet 2

1. Dr. Timothy Rowe, Professor of Biology and Geology at the University of Texas
2. He performed a CT scan.
3. The scan was performed July 29, 1999 - three months before it was to be published. Even though the CT scan revealed the fossil had serious problems, the *National Geographic* magazine was misleading when they wrote, "[*Archaeoraptor liaoningensis*] is perhaps the best evidence since *Archaeopteryx* that birds did, in fact, evolve from certain types of carnivorous dinosaurs."
4. 26 bones came from four other animals.
5. The scientist did not mention the CT scan irregularities found by Dr. Rowe. It was written up as a spectacular new "feathered dinosaur" with a two-page photo spread of the fossil and models showing what this "feathered dinosaur" looked like.

   Any Three:

   - "A Flying Dinosaur?"
   - "This fossil is perhaps the best evidence since Archaeopteryx that birds did, in fact, evolve from certain types of carnivorous dinosaurs."
   - "Preliminary studies show that this specimen has startling similarities to both dinosaurs and birds."
   - "New Birdlike Fossils are Missing Links in Dinosaur Evolution."
   - "Scientists funded by *National Geographic*... used CT scans to view parts of the animal obscured by rock. Preliminary study of the arms suggests that it was a better flier than *Archaeopteryx*, the earliest known bird. Its tail, however, is strikingly similar to the stiff tails of a family of predatory dinosaurs called *dromaeosaurs*. This mix of advanced and primitive features is exactly what scientists would expect to find in dinosaurs experimenting with flight."
6. They did a disservice in a subsequent article by whitewashing the story. The retraction implies to the reader that *National Geographic* did not know

anything about the CT scans until March, four months *after* the original story was printed. But, in fact, the scientist was told about the anomalies months *before* the original article was published in November.

7. a. Ten pages
   b. Two sentences
   c. It was put in the infrequently read *Forum* section.
8. It may have been a scientist.
9. a. They looked like any other meat-eating dinosaurs, were about six feet tall and had thick reptilian skin and large claws.
   b. They have put feathers on museum models of dinosaurs.
   c. No
10. No, they have never found feathers on the original fossils.
11. a. Ernst Haeckel's nineteenth century altering of embryo drawings.
    b. A museum scientist at the Natural History Museum in London altering human and ape fossils and planting them in the field for the discovery of a new "ape-man."
    c. Biologist Paul Kammerer faking experiments to prove adaptation.
    d. Modern scientists adding a whale's tail and flippers to *Rodhocetus.*
    e. *Archaeoraptor* altered in China and perpetuated by the *National Geographic* magazine.
    f. Feathers added to Velociraptor museum models even though they have never found feathers on the original fossils
12. If you pick and choose your artist, you could conclude that birds did evolve from dinosaurs or that birds did not evolve from dinosaurs.
13. Dr. Storrs Olson
14. a. Deinonychus
    b. Baby tyrannosaurs
    c. It is simply imaginary and has no place outside of science fiction.

Chapter 16

The Fossil Record of Flowering Plants

1. 300,000
2. 250,000
3. a. Roses
   b. Tomatoes
   c. Rhododendrons
   d. Various grasses
   e. Flowering trees such as sassafras
   f. Oak
   g. Palm
   h. Apple
4. Angiosperms
5. Even though there were plant fossils during Darwin's time, there were no fossils showing the development of the flower and its structures.
6. The poor fossil record of his day
7. Botanists continued to lament the few answers concerning plant evolution.
8. Hundreds of thousands
9. a. Flowering plants "seemed to appear suddenly during the Cretaceous period..."
   b. The no-evolution model
10. a. Microscopic pollens
    b. Delicate flowers
    c. Sepals
    d. Petals
    e. Stamens
    f. Pistils
    g. Seeds
    h. Leaves
    i. Branches
    j. Trees
11. They believe the fossils demonstrating plant evolution have yet to be collected or these plants were not fossilized.
12. They believe the lack of flowering plant ancestors speaks for itself — that plant evolution did not occur.

Chapter 17

The Origin of Life Part 1: The Formation of DNA Worksheet 1

1. About 4 billion years ago
2. A microscopic organism
3. The theoretical evolutionary event on earth when the very first form of life, a single-cell organism, formed spontaneously from chemicals.
4. This single, spontaneously formed organism mutated into all the bacteria, fungi, plants and animals that have lived on earth.
5. Nonliving chemicals
6. Yes
7. No
8. Molecules containing the element carbon
9. a. Bacteria
   b. Fungi
   c. Plant
   d. Animals (and humans)
10. a. DNA
    b. Proteins
    c. Cell membrane
11. They provide the chemical catalysts and the structure of the cell.
12. The cell membrane holds DNA and proteins together.
13. It gives instructions to the rest of the cell to make proteins, and it passes this same information on to the next generation.

The Origin of Life Part 1: The Formation of DNA Worksheet 2

1. It looks like a twisted ladder.
2. A, C, G, T
3. Three
4. A single protein containing 300 amino acids x 3 letters of DNA needed for each amino acid in this protein = 900 letter DNA strand needed for a single protein.
5. 20
6. 20 proteins needed for life to begin x 900 letters of DNA needed for each protein = 18,000 letters of DNA needed for life to begin.
7. After 20 letters coalesce, the DNA begins to break apart.
8. One out of four
9. a. 1⁄16
   b. 1⁄64
   c. 1⁄256
10. Winning the National Powerball Lottery 365 times in a row
11. When the complex, pre-assembled chemical DNA "letters" are placed in a glass beaker, the strands of DNA that coalesce are *deformed*. Specifically, they connect in the wrong "corners" of the sugar molecules which make up the DNA backbone, resulting in non-spiraled DNA.
12. Spiraling is very important because it compacts and protects the DNA.
13. They contend that if DNA (or its theoretical precursor RNA) cannot spontaneously assemble in the proper length to produce a *single* protein, the proper order to produce the needed 20 functional proteins to begin life, and the proper shape to protect the DNA from breaking up, then life could never have started from chemicals.
14. They believe such suggestions are presumptuous. Ignorance of a process does not necessarily mean it did not happen. Rather, the lack of knowledge in this area of science should spur further research and investigation.

Chapter 18

The Origin of Life Part 2: The Formation of Proteins

1. a. Produce energy
   b. Develop structures
   c. Assist in copying the DNA
2. No
3. Amino acids
4. Yes
5. Water prevents amino acids from linking together to form a protein.
6. No
7. An unnatural organic compound brought about by heating dried, purified amino acids.
8. Proteins can be thought of as a chain. A proteinoid does not look like a natural protein chain. Rather,

it looks like a bunch of chain links welded together in a clump. The links of a proteinoid are not connected properly compared to a true protein.

9. They believe proteinoids, which may have acted like proteins, came first and then eventually converted to proteins by an unknown mechanism.
10. They have suggested that the process of heating dried amino acids to form a proteinoid did not occur in the ocean but rather on the heated surface of a volcano.
11. How could significant quantities of pure, dried amino acids randomly occur in nature?
12. a. Proteinoids do not have any significant functionality, such as the ability to copy DNA or form any of the known structures in bacteria living today.
    b. They also point out that proteinoids have never been observed to form outside of a laboratory nor have they ever been observed to convert to proteins.
    c. If proteinoids cannot convert to proteins, which are part of all life forms today, how could life begin?
13. 20
14. Diseases, such as sickle cell anemia, hemophilia, and cystic fibrosis, originate from a single erroneous substitution of one amino acid for another in a protein.
15. Hundreds
16. 20
17. 6,000
18. 20 proteins needed for life to begin x 300 amino acids in each protein chain = 6,000 amino acids lined up in the correct order
19. a. If we randomly generated a new 100-amino-acid-long sequence each second, we could expect such a given enzyme to appear only once in 4 x 10 -122 years! (Convert to 10 to the 122 power.)
    b. There is not enough space in the entire universe to generate even one specific protein.
20. If proteins do not form naturally, but are necessary for life to begin, then the theory of evolution is dead.

Chapter 19

The Origin of Life Part 3: The Formation of Amino Acids Worksheet 1

1. Around 4 billion years ago
2. Amino acids
3. a. Dr. Stanley Miller
   b. 1953
4. a. Glassware
   b. Tungsten electrode
   c. Distillation
   d. Glass
5. It emulated how amino acids may have formed in nature billions of years ago.
6. A lightning bolt striking the ocean

The Origin of Life Part 3: The Formation of Amino Acids Worksheet 2

1. a. The experiment required extreme amounts of "investigator interference."
   b. He removed oxygen from his device because he knew oxygen was poisonous to the formation of amino acids.
   c. Miller removed oxygen to produce amino acids, yet oxygen is necessary to protect proteins and DNA from the sun. This would constitute a catch-22 in the model.
   d. The amino acids produced in Miller's apparatus were both right and left-handed amino acids. In contrast, nearly all living organisms today use only left-handed amino acids. Right-handed amino acids usually render a protein nonfunctional.
   e. Miller's experiment produced only a few rudimentary amino acids, not the full complement of the 20 essential amino acids that are used by living organisms today.
2. The necessary 20 amino acids needed for life to exist do not form spontaneously, without investigator interference and complex equipment.
3. Oxygen
4. If trace amounts of oxygen are present, amino acids cannot form.
5. Oxygen

6. Iron oxide minerals have been found in Greenland, dating to 3.8 billion years ago.
7. Yes
8. Miller removed oxygen to produce amino acids, yet oxygen is necessary to protect proteins and DNA from the sun.
9. Both
10. Nearly all living organisms today use only left-handed amino acids. Right-handed amino acids usually render a protein nonfunctional.
11. 20
12. No
13. The theory of evolution suggests that life may have begun in the ocean, yet water prevents the formation of proteins.
14. No
15. No

Answer Keys for Worksheets

for Use with

Living Fossils

Living Fossils Worksheet Answer Keys

Chapter 1

The Challenge That Would Change My Life

1. a. A chemical
   b. A single-cell bacterium
   c. An invertebrate like a jellyfish
   d. A fish
   e. An amphibian
   f. A reptile
   g. A mammal
   h. A monkey with a tail
   i. A tailless ape
   j. Human
2. "Ontogeny Recapitulates Phylogeny"
3. Prior to birth, animals retrace the history of evolution in their embryonic stages.
4. Extensive retouching and outrageous fudging
5. a. He made the images of different animal embryos look similar even though the embryos do not appear this way in life.
   b. He referred to neck pouches in the human embryo as "gill-arches," yet there are no fish gills in the human embryo.
   c. He referred to the end of the vertebral column of the human embryo as "a tail" even though these vertebrae coincide with the sacrum and coccyx to which the pelvic organs are attached.
6. a. 1834-1919
   b. Yes
7. a. 1977
   b. 1890s
8. a. What did I think about evolution?
   b. What did I think about the problems with the fossil record which cast doubt on the theory of evolution?
   c. What did I think about the problems with the laws of physics in the big bang model?
   d. How could life begin if proteins do not form naturally?
9. a. Nobel Prize in Physics
   b. 2004
10. Total disaster

Chapter 2

How Can You Verify Evolution?

1. a. The majority of scientists once believed the earth was the center of our planetary system, but this is false. The sun is the center.
   b. The majority also thought spontaneous generation was the explanation for how life began. This theory suggested that mice came from dirty underwear and that maggots came from rotting meat.
2. a. Too many missing links
   b. The big bang theory does not work.
   c. Life could not begin spontaneously.
3. The body of techniques used to investigate phenomena, acquire new knowledge, and verify theories
4. A scientist will start with an idea (theory or hypothesis) and then test the validity of his idea by vigorously trying to disprove it. If he or she can't falsify it, then the original theory remains tentatively true.
5. a. Animals and plants changed dramatically over time, from one type into a completely different type, through random mutations
   b. In this ever-changing line of animals and plants over millions of years, due to the survival of the fittest, the weaker predecessors became extinct.
6. a. Animals and plants would not change significantly over time.
   b. There would be fossils of modern animal and plant species in the "older" fossil layers.
7. a. Triassic
   b. Jurassic
   c. Cretaceous
8. Living fossils are fossils which look very similar to modern plants or animals.

9. a. Dragonfly
   b. Garfish
   c. Coelacanth fish
   d. Horseshoe crab
10. 108,561 miles over three continents — three times the number of miles that Charles Darwin
11. Have you found any fossils of modern animal or plant species at this dinosaur dig site?

Chapter 3

The Naming Game

1. **Dog**
   a. Kingdom: Animal
   b. Phylum: Vertebrate
   c. Class: Mammal
   d. Order: Meat-Eating
   e. Family: Dog
   f. Genus: *Canis*
   g. Species: *Familiaris*

   Horse
   h. Kingdom: Animal
   i. Phylum: Vertebrate
   j. Class: Mammal
   k. Order: Hoofed Odd Toe
   l. Family: Horse
   m. Genus: *Equus*
   n. Species: *Caballus*

   Human
   o. Kingdom: Animal
   p. Phylum: Vertebrates
   q. Class: Mammal
   r. Order: Primate
   s. Family: Apes/Human
   t. Genus: *Homo*
   u. Species: *sapien*
2. Kingdom, Phylum, Class, Order, Family, Genus, Species
3. Division
4. Genus, Species, italics
5. a. Human
   b. Horse
   c. Dog
   d. Wolf
6. A species is a group of animals that can produce fertile offspring.
7. Donkeys
8. a. 64
   b. 62
   c. 62
9. A mule is infertile and because of this, the donkey and the horse are not the same species; they are usually incapable of producing *fertile* offspring.
10. A cat cannot mate with a dog and produce offspring.
11. Tremendous
12. Species
13. a. The naming of a species is not carried out by a committee of scientists. Usually, the individual scientist who discovers the fossil assigns the species name.
    b. The scientist who discovers a fossil wholeheartedly supports evolution and his or her choice of a new species name may reflect this bias.
    c. Often, a scientist gains recognition from peers by naming a new species.
    d. There is the potential to erroneously assign a fossil a new and unique species name, even when it may look nearly the same as a modern species.
14. More than 15
15. Race
16. Similar
17. He would find modern plant and animal species in dinosaur rock layers if evolution was not valid.

Chapter 4

Echinoderms

1. a. Brittle stars

b. Sea cucumbers

c. Sea urchins

d. Sea stars

e. Sea lilies

2. Five-sided
3. a. Most sea urchins look like a tennis ball with spikes sticking out of it.

   b. A sea biscuit does not have spines.
4. a. Long-stemmed stalked crinoids

   b. Feather stars
5. a. Starfish

   b. Brittle stars

   c. Sea urchins

   d. Sea cucumbers

   e. Sea lilies
6. All five
7. Yes

Chapter 5

Aquatic Arthropods

1. a. An outer armor called an exoskeleton

   b. Segmented bodies

   c. Jointed legs
2. a. Phylum Echinodermata

   b. Phylum Arthropoda
3. a. Archaeopteryx

   b. Compsognathus
4. a. Shrimp

   b. Crayfish

   c. Fresh Water Prawns

   d. Lobsters

   e. Horseshoe crabs
5. Crabs
6. Six
7. Answers may vary but should be based on observation, examples, and facts.

Chapter 6

Land Arthropods

1. a. Insects

   b. Spiders

   c. Scorpions

   d. Millipedes

   e. Centipedes
2. 100%
3. a. Centipedes

   b. Millipedes

   c. Both of them
4. 100%
5. He would find modern land arthropods in dinosaur rock layers if evolution was not valid.
6. Names
7. Answers may vary but should be based on observation, examples, and facts.

Chapter 7

Bivalve Shellfish

1. Bivalve
2. a. Scallops

   b. Oysters

   c. Clams

   d. Mussels
3. Scallop
4. a. The shell with its ribs

   b. The hinge joint with its ears

   c. The delicate "ears" of the hinge joint tend to break off easily.
5. a. Scallops

   b. Oysters

   c. Saltwater clams

   d. Freshwater clams

   e. Mussels
6. a. Bivalve shellfish- Phylum Mollusca

   b. Snails- Phylum Mollusca

   c. Chambered nautilus- Phylum Mollusca

d. Sea cradles- Phylum Mollusca

e. Tusk shells- Phylum Mollusca

7. Yes

Chapter 8

Snails

1. a. Phylum Mollusca

   b. Phylum Mollusca

2. a. One

   b. Two

3. Dinosaur National Monument

4. a. Stegosaurus

   b. Long-necked sauropod dinosaurs

5. Petrified Forest National Park in Arizona

6. *Coelophysis*

7. Australian

8. Yes

Chapter 9

Other Types of Shellfish

1. Nautilus

2. Mollusca

3. Siphuncle

4. Elephant

5. Mollusca

6. Mollusca

7. They are marine mollusks with eight interlocking plates.

8. Dinosaur rock

9. Dinosaur rock

10. a. Bivalvia

    b. Gastropoda

    c. Cephalopoda

    d. Scaphopoda

    e. Polyplacophora

11. Living Today? Found with Dinos?

Living Today?	Found with Dinos?
Yes	Yes
Yes	Yes
Yes	Yes
Yes	Yes
Yes	Yes

12. Yes

Chapter 10

Worms

1. a. Phylum Annelida

   b. Earth worm

   c. Tube worm

2. Dinosaur, Oligochaeta

3. In the oceans

4. Answers may vary but most likely, no.

Chapter 11

Sponges and Corals

1. Porifera, Cnidaria

2. Dinosaur

3. a. Calcarea

   b. Hexactenellida

   c. Demospondgiae

4. Soft, hard

5. 34

6. Answers may vary but most likely, no.

Chapter 12

Bony Fish

1. Spinal cord

2. Backbone

3. Invertebrates

4. a. Bony fish

   b. Cartilaginous fish

   c. Jawless fish

5. Chordata

6. Bony

7. a. Sturgeon

   b. Salmon

   c. Flounder

d. Gar
e. Bowfin
f. Paddlefish
g. Eel
h. Lungfish
i. Coelacanth

8. a. Lobed-finned subclass: Coelacanth and Australian lungfish
   b. Ray-finned subclass: All other bony fish. Such as gar, sturgeon, salmon, etc.
9. Yes

Chapter 13

Cartilaginous Fish

1. Cartilaginous
2. a. Cartilaginous fish
   b. Jawless fish
   c. Bony fish
3. Chordata
4. Similar
5. Port Jackson
6. Shovelnose
7. New Mexico
8. Alberta
9. Yes
10. Angel shark, Port Jackson shark, Shovelnose ray, Goblin shark
    Modern shark teeth

Chapter 14

Jawless Fish

1. a. Lampreys
   b. Hagfish
2. Yes
3. Jawless
4. a. Bony fish
   b. Cartilaginous fish
   c. Jawless fish
5. Jawless

Chapter 15

Amphibians

1. Chordata
2. a. Fish
   b. Amphibians
   c. Reptiles
   d. Birds
   e. Mammals
3. Frogs/toads, salamanders
4. Salamander
5. Karaurus
6. Frogs
7. Yes

Chapter 16

Crocodilians

1. a. Alligators
   b. Crocodiles
   c. Gavials
2. Chordata
3. snouts
4. Alberta
5. Similar
6. Gavials
7. Germany
8. a. Alligators
   b. Crocodiles
   c. Gavials
9. Yes

Chapter 17

Snakes

1. Chordata
2. In any order:
   a. Echinodermata or echinoderms
   b. Arthropoda or arthropods
   c. Mollusca or molluscs
   d. Annelida or segmented worms

e. Porifera or sponges

f. Cnidaria or coral or Coelenterata

g. Chordata or vertebrates

3. Boa constrictor
4. Modern boa constrictor
5. Snakes

Chapter 18

Lizards

1. Chordata
2. Utah
3. Lizards
4. Ground
5. Modern
6. Rib
7. Triassic
8. Iguana

   Ground Lizard

   Tuatara

   Gliding Lizard (Draco)

Chapter 19

Turtles

1. Spinal cord
2. a. Fish

   b. Amphibians

   c. Reptiles

   d. Birds

   e. Mammals
3. a. Turtles

   b. Lizards

   c. Snakes

   d. Crocodilians
4. Dinosaur
5. Cretaceous
6. Turtle
7. Side-neck, hidden-neck
8. a. Echinoderms

   b. Arthropods

   c. Molluscs

   d. Sponges

   e. Corals

   f. Segmented worms

   g. Vertebrates
9. Modern
10. Dinosaur

Chapter 20

Birds

1. Chordata
2. a. Fish

   b. Amphibians

   c. Reptiles

   d. Birds

   e. Mammals
3. Evolution
4. a. Parrots

   b. Flamingos

   c. Cormorants

   d. Sandpipers

   e. Owls

   f. Penguins

   g. Avocets

   h. Duck-like birds

   i. Loons

   j. Albatross-like birds
5. a. *Sinornis*

   b. *Caudipteryx*

   c. *Archaeopteryx*
6. Genetic
7. Dinosaurs

Chapter 21

Mammals Worksheet 1

1. Chordata
2. a. Placental mammals
   b. Marsupial mammals
   c. Monotreme mammals
3. Absence
4. a. 1812
   b. 1812 is prior to Darwin writing *The Origin of the Species*
5. 300
6. a. Triassic
   b. Jurassic
   c. Cretaceous
7. a. Many important groups lived with the dinosaurs, not just reptiles.
   b. "Mammals are just one such important group that lived with the dinosaurs, coexisted with the dinosaurs, and survived the dinosaurs."
8. Tasmanian Devil
9. a. (possum-like) - *Gobiconodon*, 12 pounds
   b. (size of a collie) - *Repenomamus*, 30 pounds

Mammals Worksheet 2

1. a. Less than a hundred
   b. 12 or 15
2. Three
3. Hedgehog
4. Hedgehog
5. Possum
6. Possum
7. Subclasses
8. Answer should include four of the following: A small Jurassic mammal, a Triconodont; *Docodon,* a Jurassic mammal; *Sinodelphys,* a Cretaceous marsupial mammal with teeth similar to a modern opossum; *Amblotherium,* a Jurassic mammal; *Asioryctes,* a Cretaceous placental mammal; *Didelphodon,* a Cretaceous marsupial; *Priacodon,* a Jurassic mammal; *Cimolodon,* a Cretaceous mammal
9. Mesozoic

Chapter 22

Cone-Bearing Plants

1. a. Chordata (Vertebrates): any fish, amphibian, reptile, bird, or mammal
   b. Echinodermata (Five-Sided Animals): any Starfish, Brittle Stars, Sea Urchins, Sea Cucumbers, Crinoids
   c. Arthropoda (Animals with Exoskeleton): any Insects, Shrimp, Crayfish, Lobsters, Crabs, Spiders, Scorpions, Centipedes, Millipedes
   d. Mollusca (Shellfish): any Oysters, Clams, Scallops, Mussels, Snails, Nautilus, Tusk Shells, Sea Cradles
   e. Annelida (Segmented Worms): any Earthworms, Tube Worms
   f. Porifera (Sponges): any Glass Sponges, Demosponges, Bony Sponges
   g. Cnidaria (Corals): any See Pen, Humpback Coral
2. Division
3. a. Conifers
   b. Cycads
   c. Ginkgos
4. a. Cone trees are Division- Coniferae
   b. Cycads are Division- Cycadophyta
   c. Gingko trees are Division- Ginkgophyta
5. Gymnosperms
6. Sequoia
7. a. Sequoia (redwood)
   b. Dawn redwood
   c. Cook pine
   d. Bald cypress
   e. Pine trees
8. a. Cycad
   b. Coontie
9. Ginkgos
10. Dinosaur
11. Yes

Chapter 23

Spore-Forming Plants

1. a. Coniferae (cone trees): any pine tree, sequoia, redwood, bald cypress, cook pine
   b. Cycadophyta (palm-like cycads): any cycad, coontie
   c. Ginkgophyta (gingko trees): any ginkgo
2. a. Pteridophyta (ferns and horsetails): any tree fern, sensitive fern, wood fern, mosquito fern, shield fern, horsetail
   b. Lycopodiophyta (club moss): any club moss
   c. Bryophyta (moss): any peat moss
3. Dinosaurs
4. *Sphagnum*

Chapter 24

Flowering Plants

1. Magnoliophyta
2. a. Magnoliophya (flowering plants): any chestnut, oak, sassafras, poplar, walnut, ash, viburnum, magnolia, poppy, rhododendron, lily pad, sweetgum, soapberry, bay, Everglades palm
   b. Coniferae (cone trees): any pine tree, sequoia, redwood, bald cypress, cook pine
   c. Pteridophyta (Ferns and horsetails): any tree fern, sensitive fern, wood fern, mosquito fern, shield fern, horsetail
   d. Cycadophyta (palm-like cycads): any cycad, coontie
   e. Ginkgophyta (gingko trees): any ginkgo
   f. Lycopodiophyta (club moss): any club moss
   g. Bryophyta (moss): any peat moss
3. Flowering
4. Dinosaurs
5. a. Triassic
   b. Jurassic
   c. Cretaceous
6. a. Chestnut
   b. Oak
   c. Sassafras
   d. Poplar
   e. Walnut
   f. Ash
   g. Viburnum
   h. Magnolia
   i. Poppy
   j. Rhododendron
   k. Lily pad
   l. Sweetgum
   m. Soapberry
   n. Bay
   o. Everglades palm
7. Modern, Dinosaur

Chapter 25

Coming Full Circle — My Conclusions

1. Change
2. Changed, extinct
3. Dinosaur
4. Dinosaur
5. Negative evidence
6. a. Living Fossils
   b. The Origin of the Universe
   c. Traditional Adaptation Rejected by Evolution Scientists
   d. Theoretical Evolutionary Intermediates Absent for Most Organisms
   e. The Three Best Fossil Examples of Evolution Problematic
   f. Significance of Similarities Undermined
   g. Best Evidences for Evolution Eliminated Over Time
7. Displayed

Answer Keys for Tests

for Use with

Evolution: The Grand Experiment

Chapter 1

The Origin of Life: Two Opposing Views

1. One view is that an all-powerful God created the universe and all forms of life. Another view proposes that the universe began billions of years ago as a result of the big bang. Later, life in the form of a bacterium-like organism arose spontaneously from a mixture of chemicals. Subsequently, this single-cell organism slowly began to evolve into all modern life forms.
2. One of the three possible answers:
   - the evolution of whales from a land mammal
   - the evolution of birds from dinosaurs
   - the evolution of men from apes
3. One of the three possible answers:
   - Scientists have collected over one billion fossils.
   - Scientists have described the structure of DNA.
   - Scientists have identified how genes are passed on to the next generation .
4. One of the four possible answers:
   - gaps in the fossil record
   - problems with the big bang theory
   - the amazing complexity of even the simplest organisms
   - the inability of scientists to explain the origin of life using natural laws
5. God creating man
6. 10,000
7. 32 percent
8. True
9. False
10. Some educators fear teaching two opposing theories would confuse the students.

Chapter 2

Evolution's False Start: Spontaneous Generation

1. Jan Baptista von Helmont

2–4.

a. Mice came from dirty underwear.

b. Maggots came from rotting meat.

c. Bacteria (scum) came from clear boiled pond water (or meat broth).

5. 1668
6. Bacteria
7. Dr. Louis Pasteur
8. Over 2,100 years (322 B.C. + A.D. 1859 = 2,181 years)
9. 1859
10. The theory of evolution

Chapter 3

Darwin's False Mechanism for Evolution: Acquired Characteristics

1. 1859
2. The disproved idea that changes acquired in the body during life, such as enlarged muscles or a suntan, can be passed on to the next generation
3. a. Muscle building (is not passed on to the offspring).

   b. Sun tanning (is not passed on to the offspring).

   c. Stretching the neck (is not passed on to the offspring).

   d. Disuse or cutting off body parts (is not passed on to the offspring).
4. True
5. False
6. "If this implies that many parts are not modified by use and disuse during the life of the individual, I differ widely from you, as every year I come to attribute more and more to such agency."
7. The law of *use* and *disuse*.
8. August Weisman
9. August Weisman cut off the tails of mice for twenty generations in a row. (No matter how many tails he cut off, the baby mice were always born with a tail.)
10. No.

Chapter 4

Natural Selection and Chance Mutations

1. Cetacea
2. False (Breeding, either naturally or artificially, can only remove traits.)
3. Blue whale
4. (Black) bear
5. Mutations
6. Any three of the following: Hyena (-like), cat (-like), hippopotamus (-like) animals, wolf (-like), or deer (-like)
7. (Occasionally, flies would be born with) albino eyes.
8. A, C, G, and T (Any order)
9. True
10. Must have any five body changes from this list:
    - dorsal fin would have to form on back
    - bony tail would have to form a wide, cartilaginous fluke
    - body hair would have to nearly disappear
    - blubber would have to form for warmth
    - teeth would have to be replaced by a huge baleen filter
    - the nostrils would have to move from the tip of the nose to the back of the head
    - the front legs would have to change into flippers
    - body would have to increase from less than 150 pounds to 400,000 pounds
    - skull would have to increase to 14 feet long
    - back legs would have to disappear by chance
    - external ears would have to disappear
    - ears would have to allow for diving 1,650 feet deep
    - muscles for propulsion would have to change from back legs to the tail

Chapter 5

Similarities: A Basic Proof of Evolution?

1. A whale
2. A horse
3. Convergent evolution is a phenomena where two groups of organisms, possibly distantly related, evolve into similar patterns and come to look like one another.
4. Because unrelated animals also have similar body plans (or similar body parts)
5. Bear or dog (-like animals)
6. Skunk or otter (-like animals)
7. (According to evolution scientists) they evolved from different land animals that did not have finned feet; therefore, the similarities are not related to a common ancestor with finned feet.

8–10. Any three of these answers, any order:

Bird, Bat, Pterosaur (flying reptile), Flying Insects

Chapter 6

The Fossil Record and Darwin's Prediction

1. True
2. True
3. 79 percent
4. 500,000
5. True
6. 1,000
7. 1,000,000
8. Nearly one billion[1]
9. 100,000
10. 200,000

Chapter 7

The Fossil Record of Invertebrates

1. False (Invertebrates do not have a backbone.)

2-8. Insects, shellfish, starfish, trilobites, crabs, lobsters, shrimp, etc.

9. False (The Cambrian Explosion is the term used to describe the sudden appearance of invertebrate animals in the Cambrian fossil layer.)
10. False (Complicated invertebrate animals, such as trilobites, sea pens, and shellfish, appear suddenly without the predicted ancestors. Darwin's theory of

1 Older editions of *Evolution: The Grand Experiment* may have 200 million or "hundreds of millions" as the number of collected fossils. See footnote 1 in Appendix A of *Evolution: The Grand Experiment.*

evolution suggested that these ancestors would have been found once the fossil record became more complete. Since Darwin's theory was first proposed, 750,000,000 fossil invertebrates have been collected, but still, the ancestors for invertebrates, for the most part, are missing.)

Chapter 8

The Fossil Record of Fish

1. False (Vertebrates are animals *with* a backbone.)
2. False (Fossilized fish are abundant; 500,000 have been *collected* by museums.)
3. False (Fossilized fish are, in general, excellently preserved, showing even the scales, the bones, and the fins.)
4. False (The theoretical evolutionary common ancestor of all fish has not been discovered.)
5. False (The theoretical evolutionary intermediate stages between the theoretical common ancestor of all fish and the different types of fish have not been found.)
6. True
7. False
8. 750,000,000
9. 500,000
10. None

Chapter 9

The Fossil Record of Bats

1. True
2. False (The theoretical ground animal [or mammal] that evolved into a bat has not been found.)
3. 1,000
4. No
5. The fossil record would show a non-flying mammal slowly changing over millions of years of time into a bat. Intermediate animals between the ground animal (or mammal) and the bat would be evident, reflecting (for example) the development of longer finger bones to help in forming the wings, the development of enlarged muscles in the arms to move the wings, the development of a membrane attached to the wings, the development of hollow bones, etc (see page 75 of Chapter 6).
6. There would be no intermediate animals found between the theoretical ground animal (or mammal) and the first bat found in the fossil record (see page 76 of Chapter 6).
7. True
8. True
9. True
10. True

Chapter 10

The Fossil Record of Pinnipeds: Seals and Sea Lions

1. True
2. False
3. Sea lion
4. A through G
5. Does not match
6. True
7. False
8. True
9. A seal (Seals can dive to 5,200 feet and a Los Angeles class nuclear submarine can dive to only 800 feet.)
10. Two hours

Chapter 11

The Fossil Record of Flying Reptiles

1. Pterosaurs
2. a. Pterodactyls (short-tailed)
   b. *Rhamphorhynchus* (long-tailed)
3. False
4. True
5. False (Fossil pterosaurs have been found on every continent including Antarctica.)
6. True
7. True
8. None
9. None
10. 1,000

Chapter 12

The Fossil Record of Dinosaurs

1. A hundred or hundreds
2. *T. rex*
3. False (The fossil record of dinosaurs has large ancestral gaps in spite of a great number of dinosaurs found.)
4. 100,000
5. False (No evolutionary ancestors of *T. rex* have been discovered so far.)
6. 32
7. *Triceratops*
8. 700 species of dinosaurs are known.
9. No
10. None

Chapter 13

The Fossil Record of Whales

1. False (Cetaceans are a group of *aquatic* mammals.)
2. True
3. True
4. True
5. True
6. Their teeth are similar.

7–9. Hyena-like animal (*Pachyaena*), hippo-like animal and cat-like animal (*Sinonyx*). (Students should get credit for answering either the species name *or* the type of animal.)

10. No

Chapter 14

The Fossil Record of Birds — Part 1: *Archaeopteryx*

1. True
2. Germany
3. Nine
4. *Deinonychus*
5. True
6. True (Some believe dinosaurs, while others believe archosaurs.)
7. False (Dr. Wellnhofer believes *Archaeopteryx* had feathers on its head similar to a modern bird.)
8. 200,000
9. 100,000
10. False (Bats also have claws on their wings as do the extinct flying reptiles — pterosaurs.)

Chapter 15

The Fossil Record of Birds — Part 2: Feathered Dinosaurs

1. True
2. 1990s
3. 26
4. Four or five
5. True
6. True
7. False (The retraction was two sentences long in the infrequently read *Forum* section.)
8. July 1999
9. November 1999
10. False (Dr. Rowe)

Chapter 16

The Fossil Record of Flowering Plants

1. True
2. True
3. 250,000
4. Hundreds of thousands or over a million.

5–10. Any six of these, any order.

Pollen, Sepals, Petals, Stamens, Pistils, Seeds, Flowers, Leaves, Branches, Tree

Chapter 17

The Origin of Life — Part 1: The Formation of DNA

1. Around 4 billion years ago
2. True
3. a. DNA

   b. Proteins

   c. A cell membrane

4. DNA
5. Three
6. No (Long strands of DNA do not form naturally because the chemical properties of DNA prevent this. Also, *spiraled* DNA does not form naturally. Only short strands of deformed non-spiraled DNA form.)
7. Spiral or double helix or twisted ladder.
8. Origin of life
9. A, C, G, and T (any order)
10. True

Chapter 18

The Origin of Life — Part 2: The Formation of Proteins

1. True
2. Amino acids
3. No
4. Hundreds
5. Proteinoid
6. No (Even some evolution scientists are not in agreement.)
7. 20
8. Water prevents proteins from forming. Water breaks up proteins.
9. 20 necessary proteins x 300 amino acids per protein
10. False

Chapter 19

The Origin of Life — Part 3: The Formation of Amino Acids

1. Stanley Miller
2. 1953
3. Any two of the following:

   Glassware (for boiling)

   Heating device (for boiling)

   Tungsten electrode (to provide electrical charge)

   Distillation device (using a cold water condenser for separating amino acids from the tungsten electrode after they formed)

   Glassware for collecting amino acids (after distillation/condensation)
4. Two possible answers: It was an artificial device (not something found in nature) designed and built by a scientist OR the device required an extreme amount of investigator interference.
5. Both
6. Right-handed
7. True
8. Left-handed
9. No
10. Lightning

Answer Keys for Tests

for Use with

Living Fossils

Chapter 1

The Challenge That Would Change My Life

1-10. Chemical, Bacteria, Invertebrate, Fish...Amphibian, Reptile, Mammal, Monkey, Ape, Human
Give one point for each word that is in order.

Chapter 2

How Can You Verify Evolution?

1–4. a. Dragonfly
b. Garfish
c. Coelacanth fish
d. Horseshoe crab

5. Have you found any fossils of modern animal or plant species at this dinosaur dig site?

6-7.
a. He predicted that if evolution was not true, then animals and plants would not change significantly over time.
b. He predicted he should find fossils of modern animal and plant species in the older (dinosaur) fossil layers.

8-9. Any order:
a. Earth was the center of the planetary system. (The sun is.)
b. Spontaneous generation (or mice from underwear or maggots from rotting meat)

10. False. (A scientist tests a theory by trying to disprove it.)

Chapter 3

The Naming Game

Kingdom	Animal
Phylum	Vertebrates
Class	Mammal
Order	Meat-eating
Family	Dog
Genus	Canis
Species	Familiaris

Chapter 4

Echinoderms

1. Echinoderm or Echinodermata
2. False (5-fold.)
3. False (Tennis ball with spikes. Sea cucumbers look like a cucumber or banana.)

4-5. a. Stalked crinoid
b. Feather star

6-10. a. Starfish
b. Brittle stars
c. Sea urchins
d. Sea cucumbers
e. Sea lilies

Chapter 5

Aquatic Arthropods

1-5. Any five of these, any order:
- Shrimp
- Crayfish
- Fresh water prawns
- Lobsters
- Crabs
- Horseshoe crabs

6-8. Any order:
a. Outer armor (or an exoskeleton)
b. Segmented body
c. Jointed legs

9–10. Any order:
a. *Archaeopteryx*
b. *Compsognathus*

Chapter 6

Land Arthropods

1. True
2. True

3. 100 percent

4-9. Any six of these, or any examples of these, in any order: insects, centipedes, millipedes, scorpions, spiders, mites

10. True

Chapter 7

Bivalve Shellfish

1. True

2-5. Any four of these, any order: Scallops, Oysters, Clams, Saltwater clams, Freshwater clams, Mussels

6. Scallops

7-10. Any four of these, any order: Scallops, Oysters, Clams, Saltwater clams, Freshwater clams, Mussels

Chapter 8

Snails

1. False (Bivalves have two shells.)
2. True (Both belong to Phylum Mollusca)
3. True
4. True
5. True

Chapter 9

Other Types of Shellfish

1. False (Snails do not have chambers.)
2. False
3. True
4. Eight
5. False (The siphuncle is used to empty the water from the shell chambers.)
6. e
7. c
8. d
9. b
10. a

Chapter 10

Worms

1. False (Phylum Annelida)
2. False (Live in ocean)
3. True
4. True
5. True

Chapter 11

Sponges and Corals

1. False (Porifera)
2. True
3. False
4. True
5. True

6-8. Any order:

- a. Bony sponges or Calcarea
- b. Glass sponges or Hexactenellida
- c. Spongin or Demospongiae

9-10. Any order:

- a. Hard corals
- b. Soft corals

Chapter 12

Bony Fish

1-7. Any seven of these, any order: Sturgeon, coelacanth, salmon, lungfish, gar, bowfin, paddlefish, eel, flounder, or any other fish in chapter

8. Modern types of (bony) fish

9. Bony fish

10. Chordata

Chapter 13

Cartilaginous Fish

1-3. Any order: Angel Shark, Shovelnose Ray, Port Jackson Shark

4. Goblin Shark

5. True

Chapter 14

Jawless Fish

1-2. Any order: Hagfish and Lamprey

3. True
4. True
5. True

Chapter 15

Amphibians

1. False (Fossil salamander)
2. True
3. True
4. False (Phylum Chordata)
5. True

Chapter 16

Crocodilians

1. True
2. True

3-5. Any order: Alligators, Crocodiles, and Gavials

Chapter 17

Snakes

1. True
2. False
3. True

4-10. Any order: Echinodermata or echinoderms, Arthropoda or arthropods, Mollusca or molluscs, Annelida or segmented worms, Porifera or sponges, Cnidaria or coral or Coelenterata, Chordata or vertebrates

Chapter 18

Lizards

1-4. Any order: iguana-like lizards, ground lizards, tuatara-like lizards, gliding lizards

5. Chordata or vertebrates

Chapter 19

Turtles

1. True
2. True
3. True
4. True
5. True

Chapter 20

Birds

1-10. Any ten of these, any order: Parrots, flamingos, cormorants, sandpipers, owls, penguins, avocets, duck-like birds, loons, shore birds, and albatross-like birds

Chapter 21

Mammals

1. True
2. True
3. True
4. True
5. False (Nearly 100 complete skeletons have been found.)
6. True
7. True
8. True
9. False (ringtail opossum)
10. True

Chapter 22

Cone-Bearing Plants

1. Conifer or Coniferae
2. True
3. False (All seven have been found.)
4. True

5-10. Any six, any order: Sequoia or Redwood, Metasequoia or Dawn Redwood, Bald Cypress, Pine tree, Cook pine, Cycads, Coontie, Ginkgo or Maidenhair

Chapter 23

Spore-Forming Plants

1. C
2. B
3. A
4. False (all three)
5. *Sphagnum*

Chapter 24

Flowering Plants

1-10. Any ten, any order, plus any others from chapter not listed below: Rhododendrons, poppies, lily pads, sweetgum, sassafras, poplar, walnut, ash, soapberry, bay, viburnum, oak, dogwood, magnolia, chestnut, etc.

Chapter 25

Coming Full Circle — My Conclusions

1. False (less than .01%)
2. True
3. True
4. True
5. True

Sectional and Final Exams Answers

for Use with

Evolution: The Grand Experiment

Sectional Exam 1, chapters 1-3

1. One view is that an all-powerful God created the universe and all forms of life. Another view proposes that the universe began billions of years ago as a result of the big bang. Later, life in the form of a bacterium-like organism arose spontaneously from a mixture of chemicals. Subsequently, this single-celled organism slowly began to evolve into all modern life forms.
2. One of the four possible answers: Gaps in the fossil record, problems with the big bang theory, the amazing complexity of even the simplest organisms, the inability of scientists to explain the origin of life using natural laws
3. Michelangelo
4. Nearly one billion[1]
5. Processes and structures
6. The theory of evolution
7. Bacteria
8. False (Over 2,100 years: 322 B.C. + A.D. 1859 = 2,181 years)
9. True
10. True
11. The theory of spontaneous generation
12. 1859
13. The disproved idea that changes acquired in the body during life, such as enlarged muscles or a suntan, can be passed on to the next generation.
14. True
15. False (The reproductive cells [egg or sperm] are the only cells capable of passing genetic information [DNA] on to the next generation. Only changes that occur in the reproductive cells can be passed on to the next generation.)
16. a. Single-celled organism or bacterium-like organism
    b. Multicellular invertebrate
    c. Fish (or vertebrate)
    d. Amphibian
    e. Reptile
    f. Mammal
    g. Primate (ape)
    h. Man
17. Millions (or billions) of years
18. Dr. von Helmont or Dr. Jan Baptista von Helmont
19. Answer can vary. It was an idea widely accepted for two thousand years, yet it was never truly proven. In fact, Dr. Pasteur had to disprove it before scientific thinking began to change. When flawed ideas like this are accepted as fact and then form the basis of other theories, it is perpetuating bad science.
20. Scientists based their idea on horses stretching their necks to eat food, and this meant their offspring would be horses with longer necks. But using muscles to stretch its neck will not alter the horse's DNA, which is where these types of physical traits are passed to future generations.
21. Answers will vary

1 Older editions of *Evolution: The Grand Experiment* may have 200 million or "hundreds of millions" as the number of collected fossils. See footnote 1 in Appendix A of *Evolution: The Grand Experiment*.

Sectional Exam 2, chapters 4-9

1. Blue whale
2. (Black) bear
3. Mutations
4. (Occasionally, flies would be born with) albino eyes.
5. A, C, G, and T (Any order)
6. False (Porpoises, not fish)
7. False (Natural selection and artificial breeding remove traits; acting alone, they cannot add new traits.)
8. Convergent evolution is a phenomena where two groups of organisms, possibly distantly related, evolve into similar patterns and come to look like one another.
9. Because unrelated animals also have similar body plans (or similar body parts).
10. Any three of these answers, any order: bird, bat, pterosaur (flying reptile), flying insects
11. Any two of the following answers: Both have an extra thumb, both have similar teeth, both have similar skulls, both have V-shaped jaw.
12. Their ears are similar.
13. Skunk- or otter- (like animal)
14. a. Seals
    b. Sea lions

15. True
16. True
17. True
18. There would be no intermediate animals (or there would be large ancestral gaps between animals).
19. Any five of these: Fossil bacteria, embryos, eggs, jellyfish, flowers, leaves, worms, or cattails.
20. False (Invertebrates do not have a backbone.)
21. Insects, shellfish, starfish, trilobites, crabs, lobsters, shrimp, etc.
22. Ediacaran, Cambrian, Ordovician
23. True
24. False
25. Any three of the following (or others too): Sea pens, jellyfish, worms, and sponges.
26. False (The theoretical evolutionary common ancestor of all fish has not been discovered.)
27. True
28. False (Dr. Long states that the origin/evolution of sharks is a "mystery" and that "the current fossil evidence is too incomplete to answer this question.")
29. True
30. False (The theoretical ground animal [or mammal] that evolved into a bat has not been found.)
31. Answers will vary. The lack of transitional fossils is a major issue for evolutionists who say that animals changed into other animals through various processes over millions of years. Yet the lack of these fossils and the fact that some animals have remained virtually unchanged in the fossil record supports a creationist view of history which claims that animals were created.

Sectional Exam 3, chapters 10-15

1. True
2. A through G
3. Ears (External ear flap can be seen in sea lions, not seals.)
4. Sea lions
5. True
6. a. Pterodactyls (short-tailed)
   b. Rhamphorhynchus (long-tailed)
7. True
8. None
9. None
10. True
11. False (1,000 pterosaurs have been found, including fossils showing the soft-tissue membranes of the wings.)
12. False (Extinction does not prove evolution. Extinction simply means that a species or animal type no longer exists, possibly for environmental reasons.)
13. A hundred or hundreds
14. False (No evolutionary ancestors of T. rex have been discovered so far.)
15. 700 species of dinosaurs are known.
16. No
17. Ornithischian and Saurischian
18. Triceratops
19. True
20. True
21. False (Cetaceans are a group of aquatic mammals.)
22. True
23. True
24. True
25. True
26. True
27. True
28. (Black) bear
29. Hippos are plant eaters. (Yet whales are meat eaters.) Hippo fossils do not come before (not older than) whale fossils.
30. True
31. Germany
32. Nine
33. *Deinonychus*
34. True
35. True
36. False (*Deinonychus* lived after *Archaeopteryx*.)
37. True
38. True
39. 1990s
40. False (The retraction was two sentences long in the infrequently read Forum section.)
41. False (The feathers of the Chinese specimens are symmetric, suggesting to some that they were flightless birds.)

42. False (Flightless modern birds have symmetrical feathers.)
43. False (Only 12 inches or so in length)
44. True
45. True
46. Answers will vary.

Sectional Exam 4, chapters 16-19

1. True
2. True
3. 250,000
4. Hundreds of thousands or over a million
5. Roses, tomatoes, rhododendrons, the various grasses, and the flowering trees, such as sassafras, oak, palm, apple, and any of the other 250,000 flowering plants. An incorrect answer would be pine trees, conifers, cycads, or moss.
6. False (Darwin could not see the fossil evidence for flowering plants and called it a mystery.)
7. Angiosperm
8. Around 4 billion years ago
9. True
10. DNA, proteins, and a cell membrane
11. DNA
12. True
13. Winning the National Powerball Lottery every day for 365 days in a row
14. Up to twenty
15. 900
16. Spiraling makes DNA more compact and protects DNA.
17. True
18. Amino acids
19. No
20. Hundreds
21. Proteinoid
22. 20 proteins needed for life to begin x 300 amino acids in each chain
23. 20
24. The protein malfunctions or does not work or causes disease
25. True
26. Left-handed
27. False (Modern scientists have never formed a single-celled organism from chemicals in a laboratory.)
28. False (The device produced both left- and right-handed amino acids.)
29. Proteins assist in producing energy. Proteins are necessary to produce structures in the organism. Proteins assist in copying the DNA for reproduction.
30. Cell membrane, DNA, proteins (which are made of amino acids)
31. Sickle cell anemia, cystic fibrosis, hemophilia, (Or any other disease that the student can document for the teacher.)

Final Comprehensive Exam, chapters 1-19

1. One view is that an all-powerful God created the universe and all forms of life. Another view proposes that the universe began billions of years ago as a result of the big bang. Later, life in the form of a bacterium-like organism arose spontaneously from a mixture of chemicals. Subsequently, this single-cell organism slowly began to evolve into all modern life forms.
2. One of the four possible answers: gaps in the fossil record, problems with the big bang theory, the amazing complexity of even the simplest organisms, the inability of scientists to explain the origin of life using natural laws
3. Michelangelo
4. Nearly one billion[2]
5. Processes and structures
6. The theory of evolution
7. Bacteria
8. False (Over 2,100 years: 322 B.C. + A.D. 1859 = 2,181 years)
9. True
10. True
11. Any three of these answers, any order: bird, bat, pterosaur (flying reptile), flying insects
12. Any two of the following answers: Both have an extra thumb, both have similar teeth, both have similar skulls, both have V-shaped jaw.
13. Their ears are similar

2 Older editions of *Evolution: The Grand Experiment* may have 200 million or "hundreds of millions" as the number of collected fossils. See footnote 1 in Appendix A of *Evolution: The Grand Experiment.*

14. Skunk- or otter- (like animal)
15. Seals, sea lions
16. True
17. True
18. True
19. There would be no intermediate animals (or there would be large ancestral gaps between animals).
20. Any five of these: fossil bacteria, embryos, eggs, jellyfish, flowers, leaves, worms, or cattails
21. False (Invertebrates do not have a backbone.)
22. True
23. False (Cetaceans are a group of aquatic mammals.)
24. True
25. True
26. True
27. True
28. True
29. True
30. (Black) bear
31. Hippos are plant eaters. (Yet whales are meat eaters.) Hippo fossils do not come before (not older than) whale fossils.
32. True
33. Angiosperm
34. Around 4 billion years ago
35. True
36. DNA, proteins, and a cell membrane
37. DNA
38. True
39. Winning the National Powerball Lottery every day for 365 days in a row
40. Up to twenty
41. 900
42. Spiraling makes DNA more compact and protects DNA.
43. True
44. Three of the following: Observations of natural selection in action, evolution of whales from a land animal, evolution of birds from dinosaurs, evolution of men from apes
45. Three of the following: Gaps in the fossil record, problems with the big bang theory, the amazing complexity of even the simplest organisms, the inability of scientists to explain the origin of life using natural laws

Sectional and Final Exams Answers

for Use with

Living Fossils

Sectional Exam 1, chapters 1-11

1-10. Chemical...bacteria...invertebrate...fish... amphibian...reptile...mammal...monkey...ape... human

11. 1859

12-14. Any order:

a. Dr. Haeckel redrew the images of different animal embryos to make them look similar even though embryos do not appear this way in life.

b. Dr. Haeckel called the neck pouches in the human embryo "gill-arches," yet there are no fish gills in the human embryo.

c. Dr. Haeckel referred to the end of the vertebral column of the human embryo as "a tail" even though there is no tail in the human embryo.

15–18. Dragonfly, garfish, coelacanth fish, horseshoe crab

19-20. Any order:

a. Earth was the center of the planetary system. (The sun is.)

b. Spontaneous generation (or mice from underwear or maggots from rotting meat).

21-23. Any order:

a. Too many missing links.

b. The big bang theory does not work with the laws of modern physics.

c. Life could not begin spontaneously from chemicals.

24. Living fossils are fossils which look very similar to modern plants or animals.

25. True

26. True

27. False (Two animals that can produce fertile offspring.)

28. True

29. True

30-31. a. Stalked crinoid

b. Feather star

32-36. Starfish, Brittle stars, Sea urchins, Sea cucumbers, and Sea lilies

37-38. Any order:

a. Sea urchin with spikes

b. Sea urchin without spikes or sea biscuits

39. True

40-44. Any five of these, any order:

a. Shrimp

b. Crayfish

c. Fresh water prawns

d. Lobsters

e. Crabs

f. Horseshoe crabs

45–46. Any order:

a. *Archaeopteryx*

b. *Compsognathus*

47. Echinoderm or Echinodermata

48. 100 percent

49. True

50-53. Any four of these, any order:

a. Scallops

b. Oysters

c. Clams

d. Saltwater clams

e. Freshwater clams

f. Mussels

54. False

55. e

56. c

57. d

58. b

59. a

60. True

61. True

62-64. Any order:

a. Bony sponges or Calcarea

b. Glass sponges or Hexactenellida

c. Spongin or Demospongiae

65. True

Sectional Exam 2, chapters 12-25

1-3. Any three of these, any order:

Sturgeon, coelacanth, salmon, lungfish, gar, bowfin, paddlefish, eel, flounder, or any other fish in chapter.

4-6. Any order:

a. Angel shark

b. Shovelnose ray
c. Port Jackson shark

7-8. Any order: hagfish and lamprey

9. True

10. True

11. False (Phylum Chordata)

12. True

13-15. Any order: alligators, crocodiles, and gavials

16. True

17-23. Any order:

a. Echinodermata or echinoderms
b. Arthropoda or arthropods
c. Mollusca or molluscs
d. Annelida or segmented worms
e. Porifera or sponges
f. Cnidaria or coral or Coelenterata
g. Chordata or vertebrates

24-27. Any order:

a. Iguana-like lizards
b. Ground lizards
c. Tuatara-like lizards
d. Gliding lizards

28–29. Any order:

a. Ground lizard
b. Tuatara

30–33. Any order:

a. fish
b. amphibians
c. reptiles
d. birds
e. mammals

34-36. Any order:

a. *Sinornis*
b. *Caudipteryx*
c. *Archaeopteryx*

37. True

38. b

39. c

40. b

41. b

42. c

43. a or b

44. a

45-47. Any three, any order:

Ausktribosphenos nyktos or hedgehog
Gobiconodon or ringtail opossum
Opossum
Shrew
Repenomamus or Tasmanian Devil
Echidna
Duck-billed platypus
Insectivore

48-50. Any three, any order:

Sequoia or Redwood
Metasequoia or Dawn redwood
Bald cypress
Pine tree
Cook pine
Cycads
Coontie
Ginkgo or Maidenhair

51. c

52. b

53. a

54-59. Any six, any order, plus any others from chapter not listed below:

Rhododendrons, poppies, lily pads, sweetgum, sassafras, poplar, walnut, ash, soapberry, bay, viburnum, oak, dogwood, magnolia, chestnut, etc.

60. True

61-65. Any five, any order:

Living Fossils
The Origin of the Universe
The Origin of Life
Traditional Adaptation Rejected by Evolution Scientists
Theoretical Evolutionary Intermediates Absent for Most Organisms
The Three Best Fossil Examples of Evolution Problematic
Significance of Similarities Undermined
Best Evidences for Evolution Eliminated Over Time

Final Comprehensive Exam, chapters 1-25

1-10. Chemical...bacteria...invertebrate...fish... amphibian...reptile...mammal...monkey...ape... human

11-13. Any order:

a. Dr. Haeckel redrew the images of different animal embryos to make them look similar even though embryos do not appear this way in life

b. Dr. Haeckel called the neck pouches in the human embryo "gill-arches," yet there are no fish gills in the human embryo.

c. Dr. Haeckel referred to the end of the vertebral column of the human embryo as "a tail" even though there is no tail in the human embryo.

14–17. a. Dragonfly

b. Garfish

c. Coelacanth fish

d. Horseshoe crab

18-19. Any order:

a. Earth was the center of the planetary system. (The sun is.)

b. Spontaneous generation (or mice from underwear or maggots from rotting meat).

20-22. Any order:

a. Too many missing links.

b. The big bang theory does not work with the laws of modern physics.

c. Life could not begin spontaneously from chemicals.

23. Living fossils are fossils that look very similar to modern plants or animals.

24. True

25. True

26-30. a. Starfish

b. Brittle stars

c. Sea urchins

d. Sea cucumbers

e. Sea lilies

31-32. Any order:

a. Sea urchin with spikes

b. Sea urchin without spikes or sea biscuits

33-37. Any five of these, any order:

a. Shrimp

b. Crayfish

c. Fresh water prawns

d. Lobsters

e. Crabs

f. Horseshoe crabs

38–39. Any order:

a. Archaeopteryx

b. Compsognathus

40. 100 percent

41-44. Any four of these, any order: scallops, oysters, clams, saltwater clams, freshwater clams, mussels

45. False

46. e

47. c

48. d

49. b

50. a

51-53. Any three of these, any order:
Sturgeon, coelacanth, salmon, lungfish, gar, bowfin, paddlefish, eel, flounder, or any other fish in chapter.

54-56. Any order:

a. Angel shark

b. Shovelnose ray

c. Port Jackson shark

57-59. Any order:

a. Alligators

b. Crocodiles

c. Gavials

60-66. Any order:

a. Echinodermata or echinoderms

b. Arthropoda or arthropods

c. Mollusca or molluscs

d. Annelida or segmented worms

e. Porifera or sponges

f. Cnidaria or coral or Coelenterata

g. Chordata or vertebrates

67-70. Any order:

a. Iguana-like lizards

b. Ground lizards

c. Tuatara-like lizards

d. Gliding lizards

71–74. Any order:

a. Fish

b. Amphibians

c. Reptiles

d. Birds

e. Mammals

75-77. Any order:

a. *Sinornis*

b. *Caudipteryx*

c. *Archaeopteryx*

78. True

79. b

80. c

81. b

82. b

83. c

84. a or b

85. a

86-88. Any three, any order

Ausktribosphenos nyktos or hedgehog

Gobiconodon or ringtail opossum

Opossum

Shrew

Repenomamus or Tasmanian Devil

Echidna

Duck-billed platypus

Insectivore

89-91. Any three, any order:

Sequoia or redwood

Metasequoia or dawn redwood

Bald cypress

Pine tree

Cook pine

Cycads

Coontie

Ginkgo or maidenhair

92. c

93. b

94. a

95. True

96-100. Any five of these answers, any order:

Living Fossils

The Origin of the Universe

The Origin of Life

Traditional Adaptation Rejected by Evolution Scientists

Theoretical Evolutionary Intermediates Absent for Most Organisms

The Three Best Fossil Examples of Evolution Problematic

Significance of Similarities Undermined

Best Evidences for Evolution Eliminated Over Time